2004
Yearbook of
Astronomy

2004 Yearbook of Astronomy

edited by
Patrick Moore

associate editor
John Mason

MACMILLAN

First published 2003 by Macmillan
an imprint of Pan Macmillan Ltd
20 New Wharf Road, London N1 9RR
Basingstoke and Oxford
Associated companies throughout the world
www.panmacmillan.com

ISBN 0 333 98941 4

Copyright © Macmillan Publishers 2003

9 8 7 6 5 4 3 2

A CIP catalogue record for this book is available from
the British Library.

Typeset by Rowland Phototypesetting Ltd,
Bury St Edmunds, Suffolk
Printed and bound in Great Britain by
Mackays of Chatham plc, Chatham, Kent

Contents

Part III
Miscellaneous

Editor's Foreword

The latest *Yearbook* follows the traditional pattern, but with one important change: this year we have included star charts to help you find Uranus and Neptune. Undoubtedly, two of the highlights of 2004 will be the long-awaited transit of Venus on June 8, and the arrival of the Cassini–Huygens spacecraft at Saturn early the following month. These events both feature in our articles section, where we have contributions both from our regular authors, notably Paul Murdin, Chris Kitchin, Fred Watson and Allan Chapman, and from some very welcome newcomers. As usual, we have done our best to cover a wide range, both of subject and of technical level. Gordon Taylor has, as always, produced the material for the monthly notes, Bob Argyle has provided the information on double stars, and John Isles has completely updated the section dealing with variable stars. I am also delighted that Dr John Mason again joins me as Associate Editor. It is a pleasure to be working with him.

PATRICK MOORE
Selsey, August 2003

Preface

New readers will find that all the information in this *Yearbook* is given in diagrammatic or descriptive form; the positions of the planets may easily be found from the specially designed star charts, while the monthly notes describe the movements of the planets and give details of other astronomical phenomena visible in both the Northern and Southern Hemispheres. Two sets of star charts are provided. The **Northern Charts** (pp. 17 to 41) are designed for use at latitude 52°N, but may be used without alteration throughout the British Isles, and (except in the case of eclipses and occultations) in other countries of similar northerly latitude. The **Southern Charts** (pp. 43 to 67) are drawn for latitude 35°S, and are suitable for use in South Africa, Australia and New Zealand, and other locations in approximately the same southerly latitude. The reader who needs more detailed information will find *Norton's Star Atlas* an invaluable guide, while more precise positions of the planets and their satellites, together with predictions of occultations, meteor showers and periodic comets, may be found in the *Handbook* of the British Astronomical Association. Readers will also find details of forthcoming events given in the American monthly magazine *Sky & Telescope* and the British periodical *Astronomy Now*.

Important Note

The times given on the star charts and in the Monthly Notes are generally given as local times, using the 24-hour clock, the day beginning at midnight. All the dates, and the times of a few events (e.g. eclipses), are given in Greenwich Mean Time (GMT), which is related to local time by the formula

Local Mean Time = GMT − west longitude

In practice, small differences in longitude are ignored, and the observer will use local clock time, which will be the appropriate Standard (or Zone) Time. As the formula indicates, places in west longitude will have a Standard Time slow on GMT, while places in east longitude will have a Standard Time fast on GMT. As examples we have:

Standard Time in

New Zealand	GMT + 12 hours
Victoria, NSW	GMT + 10 hours
Western Australia	GMT + 8 hours
South Africa	GMT + 2 hours
British Isles	GMT
Eastern ST	GMT − 5 hours
Central ST	GMT − 6 hours, etc.

If Summer Time is in use, the clocks will have been advanced by one hour, and this hour must be subtracted from the clock time to give Standard Time.

Part I

Monthly Charts and Astronomical Phenomena

Notes on the Star Charts

The stars, together with the Sun, Moon and planets, seem to be set on the surface of the celestial sphere, which appears to rotate about the Earth from east to west. Since it is impossible to represent a curved surface accurately on a plane, any kind of star map is bound to contain some form of distortion.

Most of the monthly star charts which appear in the various journals and some national newspapers are drawn in circular form. This is perfectly accurate, but it can make the charts awkward to use. For the star charts in this volume, we have preferred to give two hemispherical maps for each month of the year, one showing the northern aspect of the sky and the other showing the southern aspect. Two sets of monthly charts are provided, one for observers in the Northern Hemisphere and one for those in the Southern Hemisphere.

Unfortunately the constellations near the overhead point (the zenith) on these hemispherical charts can be rather distorted. This would be a serious drawback for precision charts, but what we have done is to give maps which are best suited to star recognition. We have also refrained from putting in too many stars, so that the main patterns stand out clearly. To help observers with any distortions near the zenith, and the lack of overlap between the charts of each pair, we have also included two circular maps, one showing all the constellations in the northern half of the sky, and one those in the southern half. Incidentally, there is a curious illusion that stars at an altitude of 60° or more are actually overhead, and beginners may often feel that they are leaning over backwards in trying to see them.

The charts show all stars down to the fourth magnitude, together with a number of fainter stars which are necessary to define the shapes of constellations. There is no standard system for representing the outlines of the constellations, and triangles and other simple figures have been used to give outlines which are easy to trace with the naked eye. The names of the constellations are given, together with the proper names of the brighter stars. The apparent magnitudes of the stars are indicated roughly by using different sizes of dot, the larger dots representing the brighter stars.

The two sets of star charts – one each for Northern and Southern Hemisphere observers – are similar in design. At each opening there is a single circular chart which shows all the constellations in that hemisphere of the sky. (These two charts are centred on the North and South Celestial Poles, respectively.) Then there are twelve double-page spreads, showing the northern and southern aspects for each month of the year for observers in that hemisphere. In the **Northern Charts** (drawn for latitude 52°N) the left-hand chart of each spread shows the northern half of the sky (lettered 1N, 2N, 3N . . . 12N), and the corresponding right-hand chart shows the southern half of the sky (lettered 1S, 2S, 3S . . . 12S). The arrangement and lettering of the charts is exactly the same for the **Southern Charts** (drawn for latitude 35°S).

Because the sidereal day is shorter than the solar day, the stars appear to rise and set about four minutes earlier each day, and this amounts to two hours in a month. Hence the twelve pairs of charts in each set are sufficient to give the appearance of the sky throughout the day at intervals of two hours, or at the same time of night at monthly intervals throughout the year. For example, charts 1N and 1S here are drawn for 23 hours on January 6. The view will also be the same on October 6 at 05 hours; November 6 at 03 hours; December 6 at 01 hours and February 6 at 21 hours. The actual range of dates and times when the stars on the charts are visible is indicated on each page. Each pair of charts is numbered in bold type, and the number to be used for any given month and time may be found from the following table:

Local Time	18h	20h	22h	0h	02h	04h	06h
January	11	12	1	2	3	4	5
February	12	1	2	3	4	5	6
March	1	2	3	4	5	6	7
April	2	3	4	5	6	7	8
May	3	4	5	6	7	8	9
June	4	5	6	7	8	9	10
July	5	6	7	8	9	10	11
August	6	7	8	9	10	11	12
September	7	8	9	10	11	12	1
October	8	9	10	11	12	1	2

Local Time	18h	20h	22h	0h	02h	04h	06h
November	9	10	11	12	1	2	3
December	10	11	12	1	2	3	4

On these charts, the ecliptic is drawn as a broken line on which longitude is marked every 10°. The positions of the planets are then easily found by reference to the table on p. 74. It will be noticed that on the **Southern Charts** the ecliptic may reach an altitude in excess of 62½° on the star charts showing the northern aspect (5N to 9N). The continuations of the broken line will be found on the corresponding charts for the southern aspect (5S, 6S, 8S and 9S).

Northern Star Charts

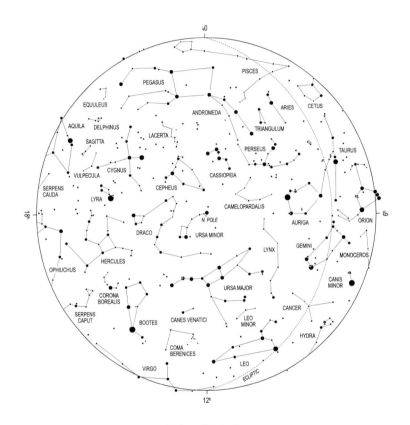

Northern Hemisphere

Note that the markers at 0ʰ, 6ʰ, 12ʰ and 18ʰ
indicate hours of Right Ascension.

1N

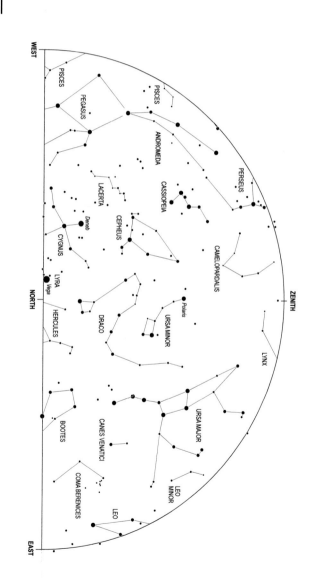

October 6 at 5h
November 6 at 3h
December 6 at 1h
January 6 at 23h
February 6 at 21h

October 21 at 4h
November 21 at 2h
December 21 at midnight
January 21 at 22h
February 21 at 20h

1S

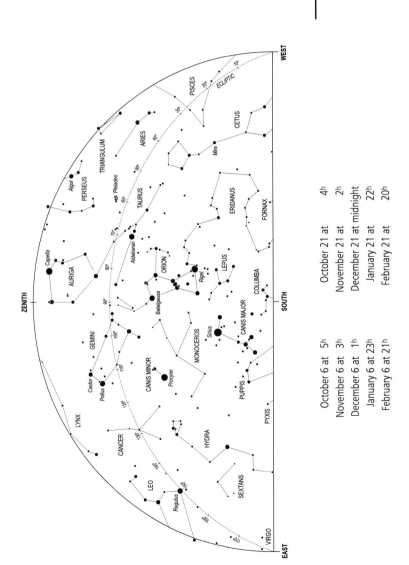

October 21 at	4ʰ
November 21 at	2ʰ
December 21 at midnight	
January 21 at	22ʰ
February 21 at	20ʰ

October 6 at	5ʰ
November 6 at	3ʰ
December 6 at	1ʰ
January 6 at	23ʰ
February 6 at	21ʰ

2N

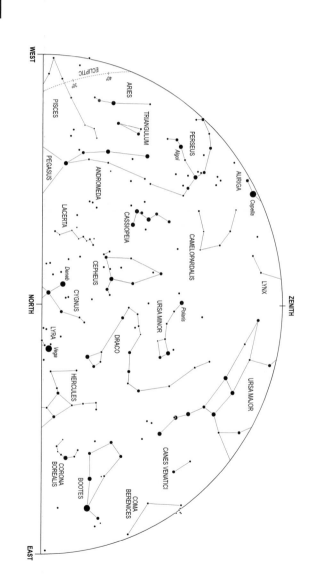

November 6 at 5h
December 6 at 3h
January 6 at 1h
February 6 at 23h
March 6 at 21h

November 21 at 4h
December 21 at 2h
January 21 at midnight
February 21 at 22h
March 21 at 20h

2S

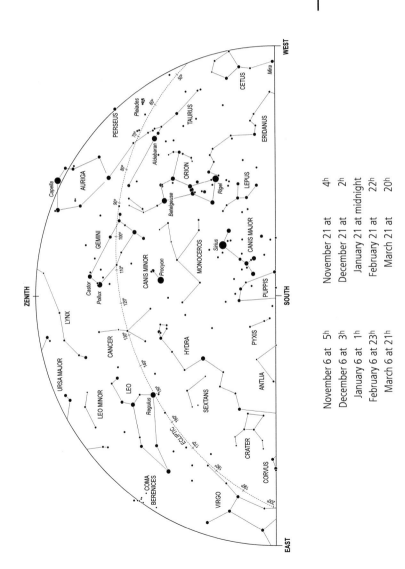

November 21 at	4ʰ
December 21 at	2ʰ
January 21 at midnight	
February 21 at	22ʰ
March 21 at	20ʰ

November 6 at	5ʰ
December 6 at	3ʰ
January 6 at	1ʰ
February 6 at	23ʰ
March 6 at	21ʰ

3N

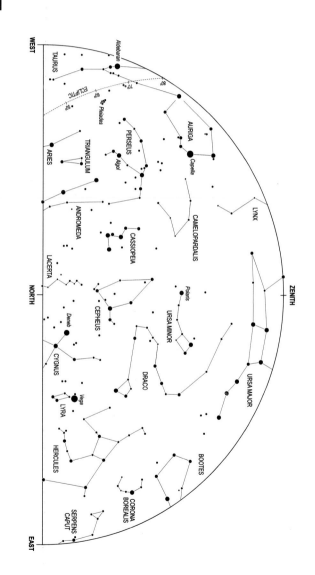

December 6 at 5h
January 6 at 3h
February 6 at 1h
March 6 at 23h
April 6 at 21h

December 21 at 4h
January 21 at 2h
February 21 at midnight
March 21 at 22h
April 21 at 20h

3S

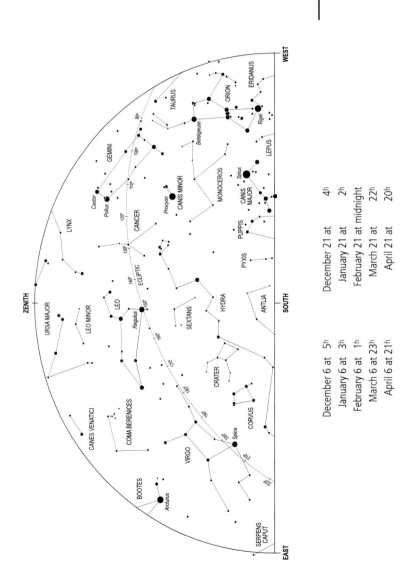

December 21 at	4ʰ
January 21 at	2ʰ
February 21 at midnight
March 21 at	22ʰ
April 21 at	20ʰ

December 6 at	5ʰ
January 6 at	3ʰ
February 6 at	1ʰ
March 6 at	23ʰ
April 6 at	21ʰ

4N

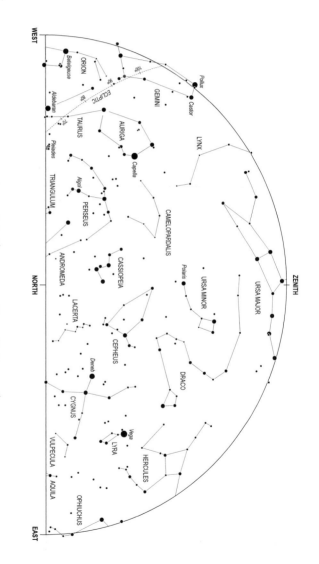

January 6 at 5h
February 6 at 3h
March 6 at 1h
April 6 at 23h
May 6 at 21h

January 21 at 4h
February 21 at 2h
March 21 at midnight
April 21 at 22h
May 21 at 20h

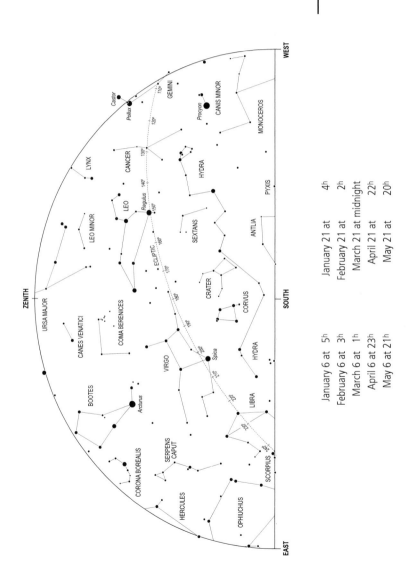

January 21 at   4h
February 21 at   2h
March 21 at midnight
April 21 at   22h
May 21 at   20h

January 6 at 5h
February 6 at 3h
March 6 at   1h
April 6 at 23h
May 6 at 21h

5N

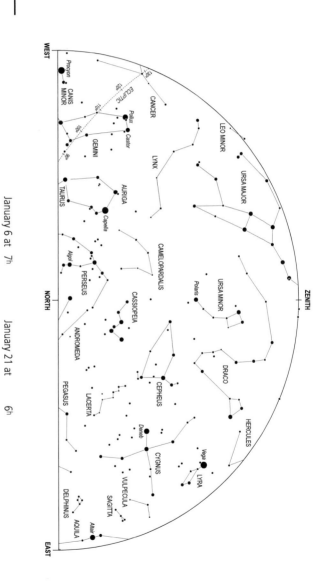

January 6 at 7h
February 6 at 5h
March 6 at 3h
April 6 at 1h
May 6 at 23h

January 21 at 6h
February 21 at 4h
March 21 at 2h
April 21 at midnight
May 21 at 22h

5S

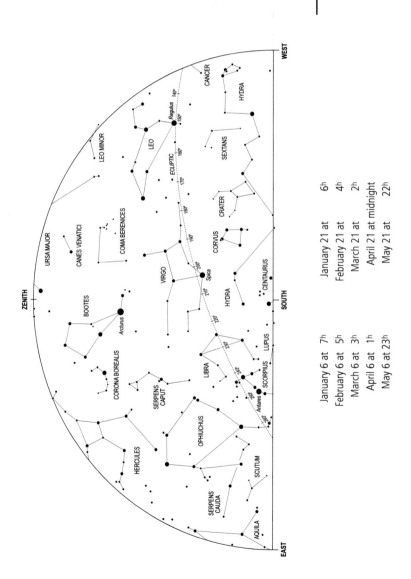

January 6 at	7ʰ	January 21 at	6ʰ
February 6 at	5ʰ	February 21 at	4ʰ
March 6 at	3ʰ	March 21 at	2ʰ
April 6 at	1ʰ	April 21 at midnight
May 6 at	23ʰ	May 21 at	22ʰ

6N

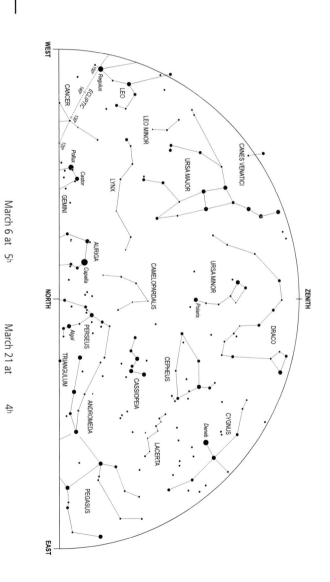

March 6 at 5h
April 6 at 3h
May 6 at 1h
June 6 at 23h
July 6 at 21h

March 21 at 4h
April 21 at 2h
May 21 at midnight
June 21 at 22h
July 21 at 20h

6S

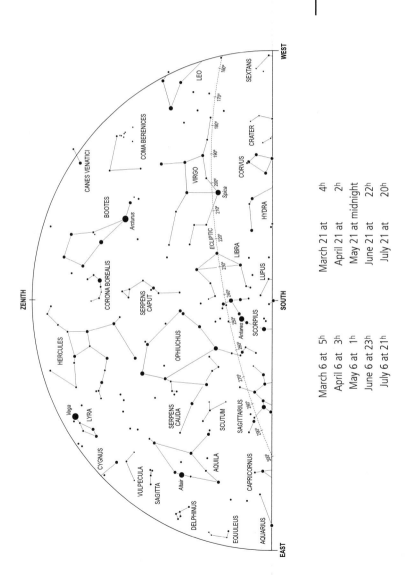

WEST

ZENITH

SOUTH

EAST

LEO
160°
170°
SEXTANS
COMA BERENICES
180°
CANES VENATICI
190°
CRATER
VIRGO
200°
CORVUS
BOOTES
Spica
210°
Arcturus
ECLIPTIC
220°
HYDRA
CORONA BOREALIS
230°
LIBRA
SERPENS
CAPUT
240°
LUPUS
HERCULES
250°
Antares
SCORPIUS
OPHIUCHUS
260°
Vega
270°
LYRA
SERPENS
CAUDA
280°
SCUTUM
CYGNUS
290°
SAGITTARIUS
VULPECULA
AQUILA
300°
SAGITTA
Altair
CAPRICORNUS
DELPHINUS
EQUULEUS
AQUARIUS

March 21 at   4h
April 21 at   2h
May 21 at midnight
June 21 at   22h
July 21 at   20h

March 6 at   5h
April 6 at   3h
May 6 at   1h
June 6 at 23h
July 6 at 21h

7N

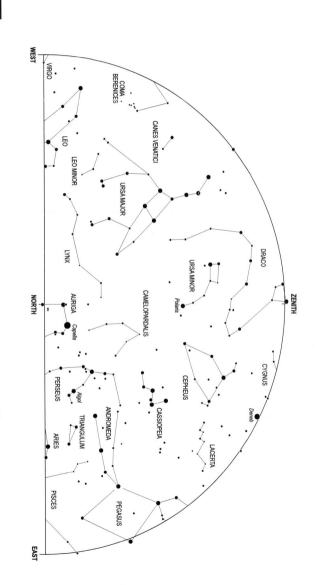

May 6 at 3h
June 6 at 1h
July 6 at 23h
August 6 at 21h
September 6 at 19h

May 21 at 2h
June 21 at midnight
July 21 at 22h
August 21 at 20h
September 21 at 18h

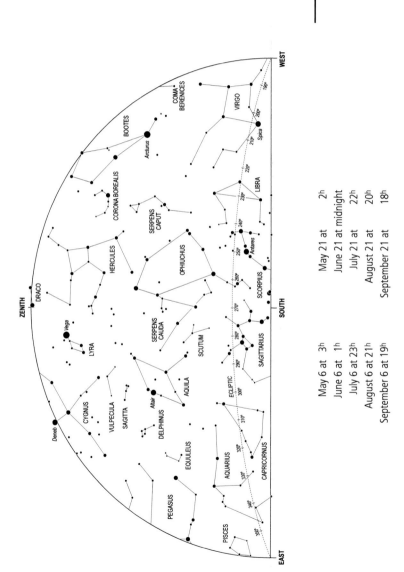

May 21 at 2ʰ
June 21 at midnight
July 21 at 22ʰ
August 21 at 20ʰ
September 21 at 18ʰ

May 6 at 3ʰ
June 6 at 1ʰ
July 6 at 23ʰ
August 6 at 21ʰ
September 6 at 19ʰ

8N

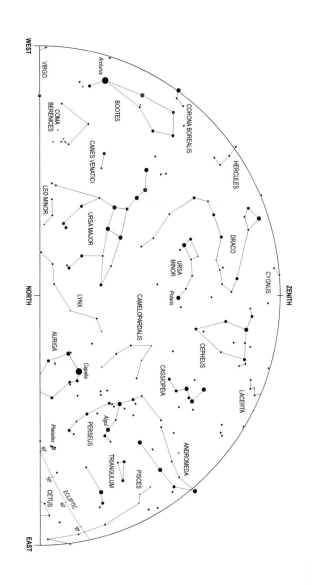

July 6 at 1h
August 6 at 23h
September 6 at 21h
October 6 at 19h
November 6 at 17h

July 21 at midnight
August 21 at 22h
September 21 at 20h
October 21 at 18h
November 21 at 16h

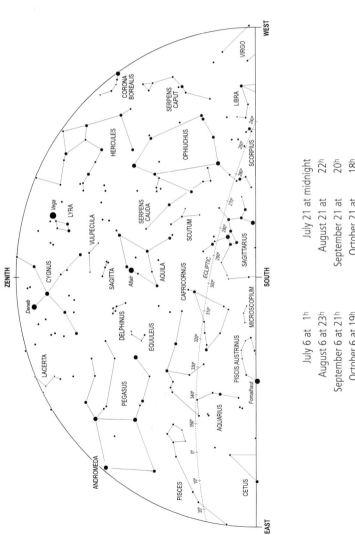

July 21 at midnight
August 21 at 22h
September 21 at 20h
October 21 at 18h
November 21 at 16h

July 6 at 1h
August 6 at 23h
September 6 at 21h
October 6 at 19h
November 6 at 17h

9N

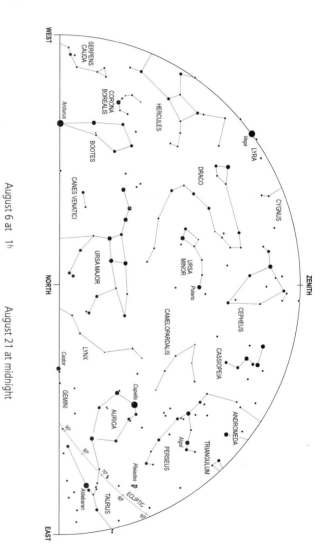

August 6 at 1h
September 6 at 23h
October 6 at 21h
November 6 at 19h
December 6 at 17h

August 21 at midnight
September 21 at 22h
October 21 at 20h
November 21 at 18h
December 21 at 16h

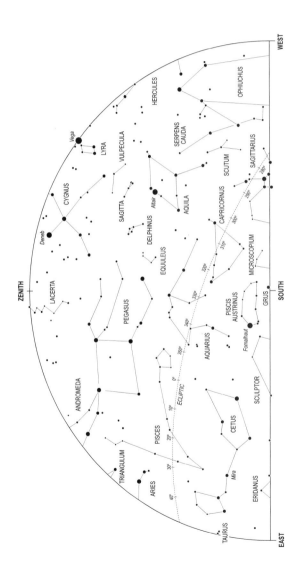

August 6 at 1ʰ   August 21 at midnight
September 6 at 23ʰ   September 21 at 22ʰ
October 6 at 21ʰ   October 21 at 20ʰ
November 6 at 19ʰ   November 21 at 18ʰ
December 6 at 17ʰ   December 21 at 16ʰ

10N

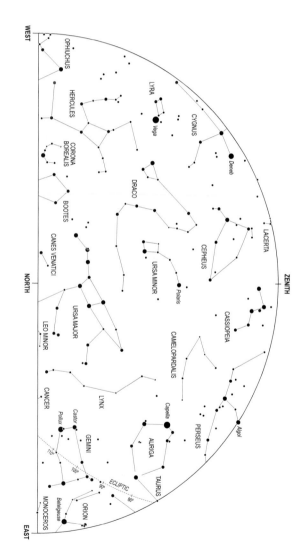

WEST

OPHIUCHUS
HERCULES
LYRA
Vega
CYGNUS
Deneb
CORONA BOREALIS
DRACO
LACERTA
BOOTES
CEPHEUS
CANES VENATICI
URSA MINOR
Polaris
CASSIOPEIA
NORTH
URSA MAJOR
CAMELOPARDALIS
LEO MINOR
CANCER
LYNX
Capella
PERSEUS
Castor
Pollux
GEMINI
AURIGA
Algol
110°
100°
90°
80°
ECLIPTIC
TAURUS
MONOCEROS
Betelgeuse
ORION

ZENITH

EAST

August 6 at   3ʰ
September 6 at   1ʰ
October 6 at   23ʰ
November 6 at   21ʰ
December 6 at   19ʰ

August 21 at   2ʰ
September 21 at midnight
October 21 at   22ʰ
November 21 at   20ʰ
December 21 at   18ʰ

10S

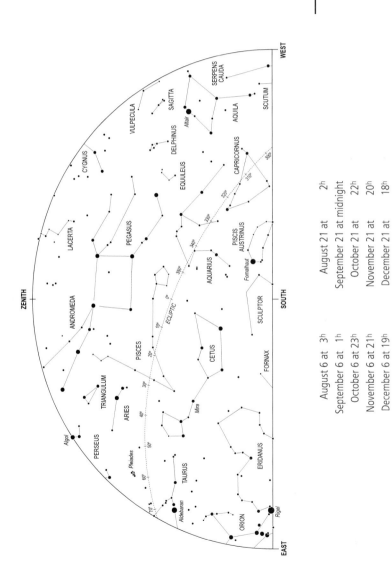

ZENITH

WEST

SERPENS
CAUDA

SAGITTA

SCUTUM

VULPECULA

Altair

AQUILA

DELPHINUS

CYGNUS

CAPRICORNUS

300°

EQUULEUS

310°

LACERTA

320°

PEGASUS

330°

PISCIS
AUSTRINUS

340°

Fomalhaut

ANDROMEDA

350°

AQUARIUS

SCULPTOR

0°

SOUTH

ECLIPTIC

10°

PISCES

CETUS

FORNAX

20°

TRIANGULUM

30°

ARIES

Mira

Algol

40°

PERSEUS

Pleiades

50°

TAURUS

ERIDANUS

60°

Aldebaran

70°

ORION

Rigel

EAST

August 6 at   3ʰ
September 6 at   1ʰ
October 6 at 23ʰ
November 6 at 21ʰ
December 6 at 19ʰ

August 21 at   2ʰ
September 21 at midnight
October 21 at 22ʰ
November 21 at 20ʰ
December 21 at 18ʰ

11N

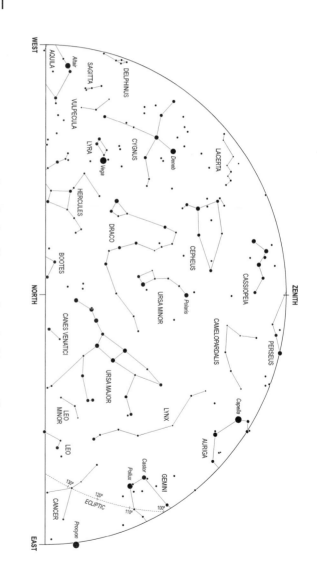

September 6 at 3h
October 6 at 1h
November 6 at 23h
December 6 at 21h
January 6 at 19h

September 21 at   2h
October 21 at midnight
November 21 at   22h
December 21 at   20h
January 21 at   18h

11S

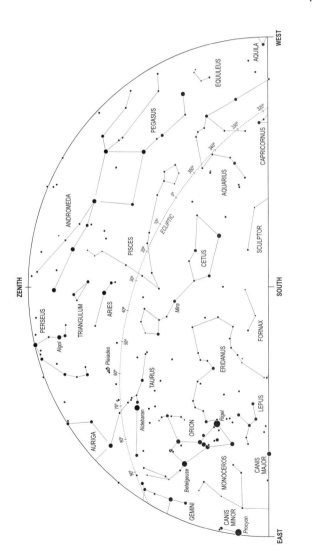

September 6 at 3ʰ
October 6 at 1ʰ
November 6 at 23ʰ
December 6 at 21ʰ
January 6 at 19ʰ

September 21 at 2ʰ
October 21 at midnight
November 21 at 22ʰ
December 21 at 20ʰ
January 21 at 18ʰ

12N

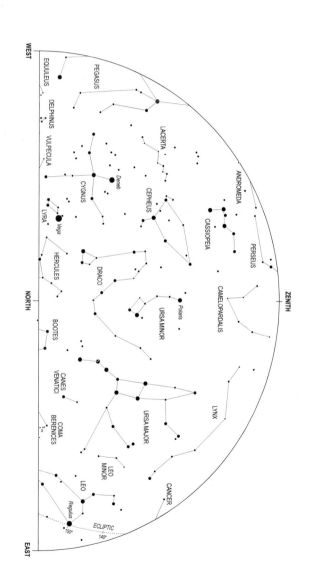

October 6 at 3h
November 6 at 1h
December 6 at 23h
January 6 at 21h
February 6 at 19h

October 21 at 2h
November 21 at midnight
December 21 at 22h
January 21 at 20h
February 21 at 18h

12S

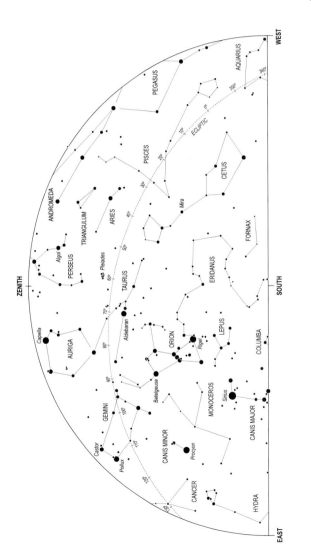

October 21 at 2ʰ
November 21 at midnight
December 21 at 22ʰ
January 21 at 20ʰ
February 21 at 18ʰ

October 6 at 3ʰ
November 6 at 1ʰ
December 6 at 23ʰ
January 6 at 21ʰ
February 6 at 19ʰ

Southern Star Charts

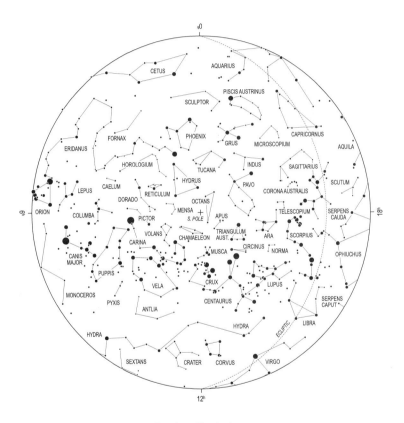

Southern Hemisphere

Note that the markers at 0ʰ, 6ʰ, 12ʰ and 18ʰ
indicate hours of Right Ascension.

1N

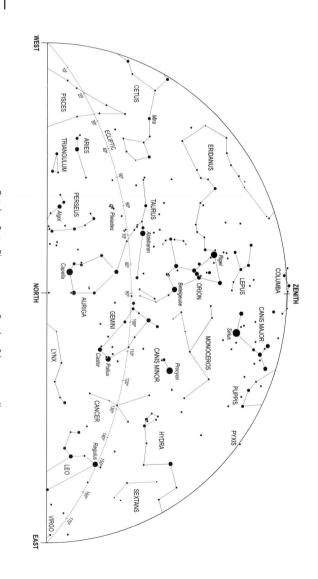

October 6 at   5h
November 6 at   3h
December 6 at   1h
January 6 at 23h
February 6 at 21h

October 21 at   4h
November 21 at   2h
December 21 at midnight
January 21 at   22h
February 21 at   20h

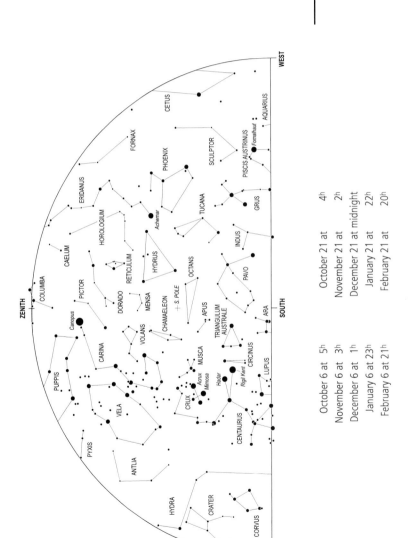

October 6 at 5ʰ	October 21 at 4ʰ
November 6 at 3ʰ	November 21 at 2ʰ
December 6 at 1ʰ	December 21 at midnight
January 6 at 23ʰ	January 21 at 22ʰ
February 6 at 21ʰ	February 21 at 20ʰ

2N

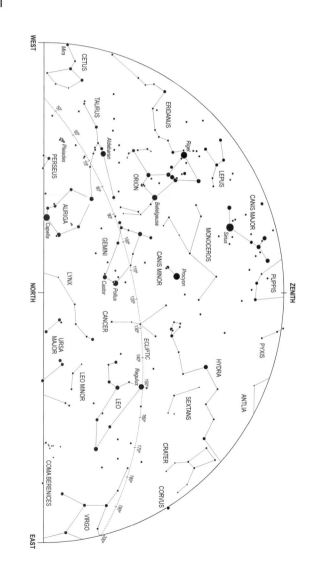

WEST

NORTH

EAST

ZENITH

November 6 at   5h
December 6 at   3h
January 6 at   1h
February 6 at 23h
March 6 at 21h

November 21 at   4h
December 21 at   2h
January 21 at midnight
February 21 at   22h
March 21 at   20h

Mira
CETUS
TAURIS
PERSEUS
Pleiades
Aldebaran
AURIGA
Capella
ERIDANUS
ORION
Rigel
Betelgeuse
LEPUS
CANIS MAJOR
Sirius
MONOCEROS
PUPPIS
PYXIS
ANTLIA
GEMINI
Castor
Pollux
LYNX
CANIS MINOR
Procyon
CANCER
URSA MAJOR
LEO MINOR
LEO
Regulus
HYDRA
SEXTANS
CRATER
CORVUS
COMA BERENICES
VIRGO
ECLIPTIC

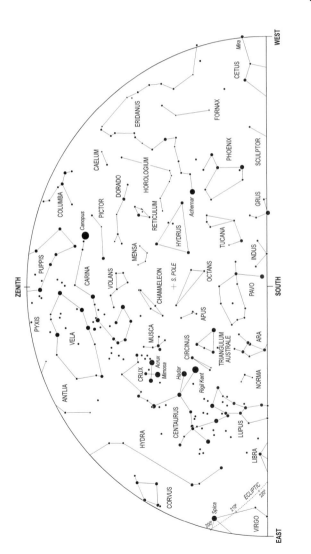

November 21 at 4ʰ
December 21 at 2ʰ
January 21 at midnight
February 21 at 22ʰ
March 21 at 20ʰ

November 6 at 5ʰ
December 6 at 3ʰ
January 6 at 1ʰ
February 6 at 23ʰ
March 6 at 21ʰ

3N

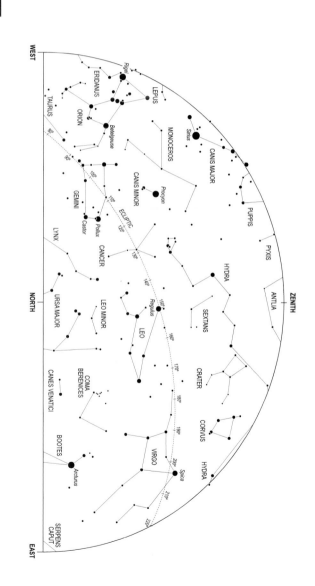

WEST

NORTH

EAST

ZENITH

January 6 at 3h
February 6 at 1h
March 6 at 23h
April 6 at 21h
May 6 at 19h

January 21 at 2h
February 21 at midnight
March 21 at 22h
April 21 at 20h
May 21 at 18h

3S

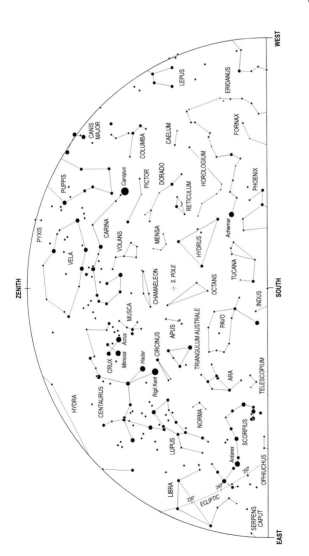

ZENITH

WEST

SOUTH

EAST

PYXIS
PUPPIS
CANIS MAJOR
Canopus
VELA
CARINA
VOLANS
COLUMBA
LEPUS
CAELUM
PICTOR
DORADO
RETICULUM
HOROLOGIUM
ERIDANUS
FORNAX
PHOENIX
Achernar
HYDRUS
TUCANA
MENSA
CHAMAELEON
+ S. POLE
OCTANS
INDUS
Acrux
Mimosa
Hadar
MUSCA
CRUX
CIRCINUS
APUS
Rigil Kent
TRIANGULUM AUSTRALE
PAVO
CENTAURUS
HYDRA
NORMA
ARA
TELESCOPIUM
LUPUS
SCORPIUS
Antares
LIBRA
230°
240°
250°
ECLIPTIC
OPHIUCHUS
SERPENS CAPUT

January 21 at   2ʰ
February 21 at midnight
March 21 at   22ʰ
April 21 at   20ʰ
May 21 at   18ʰ

January 6 at   3ʰ
February 6 at   1ʰ
March 6 at 23ʰ
April 6 at 21ʰ
May 6 at 19ʰ

4N

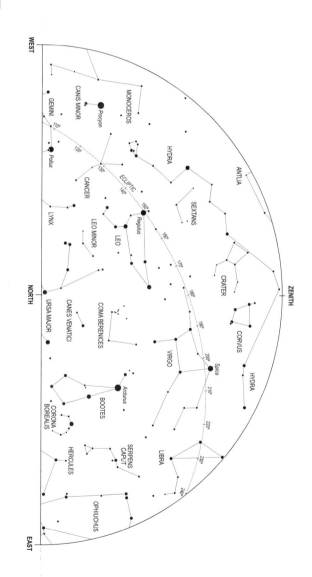

February 6 at 3h
March 6 at 1h
April 6 at 23h
May 6 at 21h
June 6 at 19h

February 21 at 2h
March 21 at midnight
April 21 at 22h
May 21 at 20h
June 21 at 18h

WEST
NORTH
EAST
ZENITH

MONOCEROS
CANIS MINOR
Procyon
GEMINI
Pollux
CANCER
LYNX
LEO MINOR
LEO
Regulus
SEXTANS
HYDRA
ANTLIA
CRATER
CORVUS
HYDRA
Spica
VIRGO
COMA BERENICES
CANES VENATICI
URSA MAJOR
Arcturus
BOOTES
CORONA BOREALIS
HERCULES
SERPENS CAPUT
LIBRA
OPHIUCHUS

ECLIPTIC

110° 120° 130° 140° 150° 160° 170° 180° 190° 200° 210° 220° 230° 240°

4S

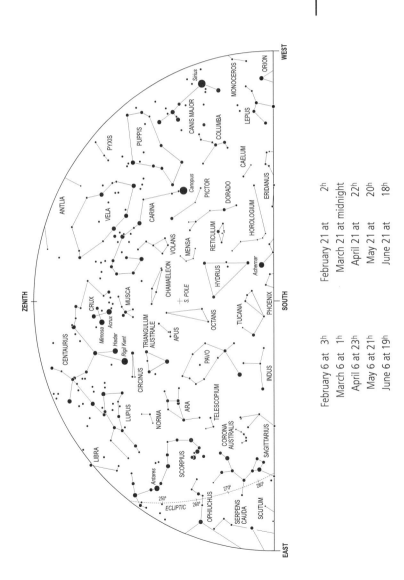

ZENITH

WEST

ORION
MONOCEROS
CANIS MAJOR
Sirius
COLUMBA
LEPUS
PYXIS
PUPPIS
CAELUM
ANTLIA
Canopus
PICTOR
ERIDANUS
VELA
CARINA
DORADO
HOROLOGIUM
VOLANS
MENSA
RETICULUM
CHAMAELEON
HYDRUS
S. POLE
Achernar
CRUX
MUSCA
OCTANS
PHOENIX
Mimosa
Acrux
TUCANA
CENTAURUS
Hadar
APUS
Rigil Kent
TRIANGULUM
AUSTRALE
SOUTH
CIRCINUS
PAVO
LUPUS
NORMA
ARA
INDUS
LIBRA
TELESCOPIUM
Antares
SCORPIUS
CORONA
AUSTRALIS
SAGITTARIUS
250°
270°
280°
ECLIPTIC
260°
OPHIUCHUS
SERPENS
CAUDA
SCUTUM

EAST

| February 21 at | 2ʰ |
| March 21 at midnight |
April 21 at	22ʰ
May 21 at	20ʰ
June 21 at	18ʰ

| February 6 at | 3ʰ |
| March 6 at | 1ʰ |
| April 6 at 23ʰ |
| May 6 at 21ʰ |
| June 6 at 19ʰ |

5N

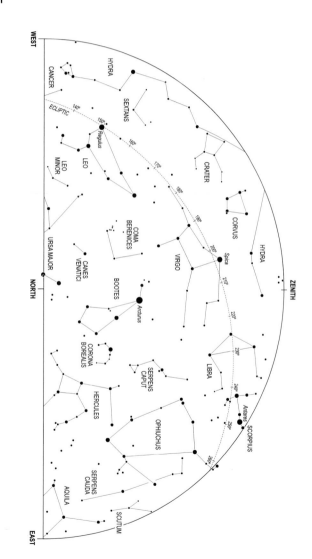

March 6 at 3ʰ
April 6 at 1ʰ
May 6 at 23ʰ
June 6 at 21ʰ
July 6 at 19ʰ

March 21 at 2ʰ
April 21 at midnight
May 21 at 22ʰ
June 21 at 20ʰ
July 21 at 18ʰ

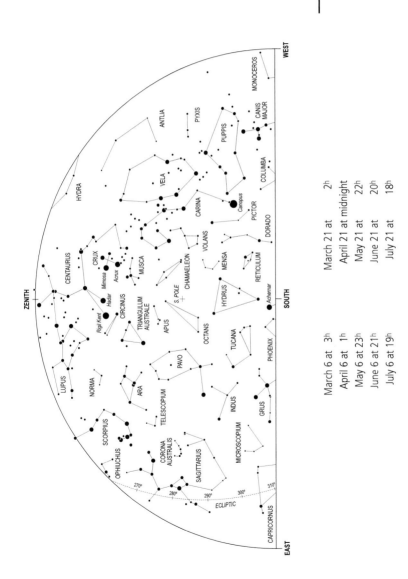

March 21 at 2ʰ
April 21 at midnight
May 21 at 22ʰ
June 21 at 20ʰ
July 21 at 18ʰ

March 6 at 3ʰ
April 6 at 1ʰ
May 6 at 23ʰ
June 6 at 21ʰ
July 6 at 19ʰ

6N

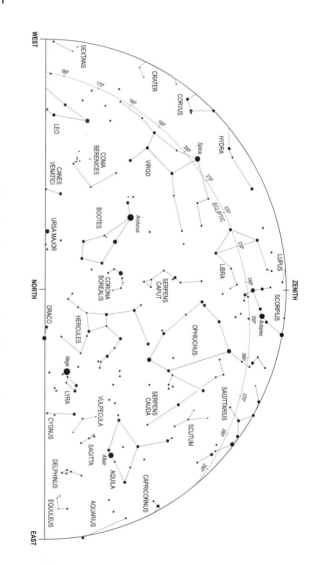

March 6 at 5h
April 6 at 3h
May 6 at 1h
June 6 at 23h
July 6 at 21h

March 21 at 4h
April 21 at 2h
May 21 at midnight
June 21 at 22h
July 21 at 20h

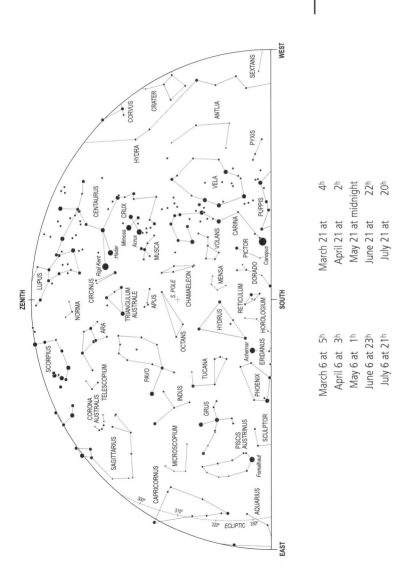

WEST

ZENITH

SOUTH

EAST

March 21 at	4ʰ
April 21 at	2ʰ
May 21 at midnight	
June 21 at	22ʰ
July 21 at	20ʰ

March 6 at	5ʰ
April 6 at	3ʰ
May 6 at	1ʰ
June 6 at	23ʰ
July 6 at	21ʰ

7N

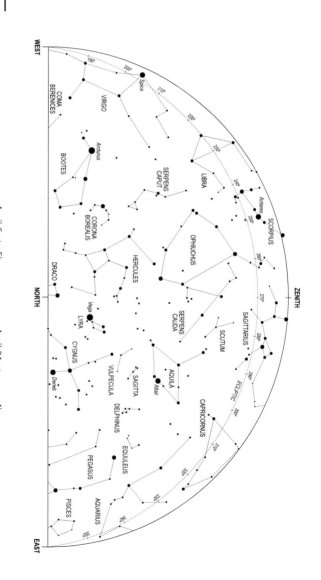

April 6 at	5h	April 21 at	4h
May 6 at	3h	May 21 at	2h
June 6 at	1h	June 21 at midnight
July 6 at	23h	July 21 at	22h
August 6 at	21h	August 21 at	20h

7S

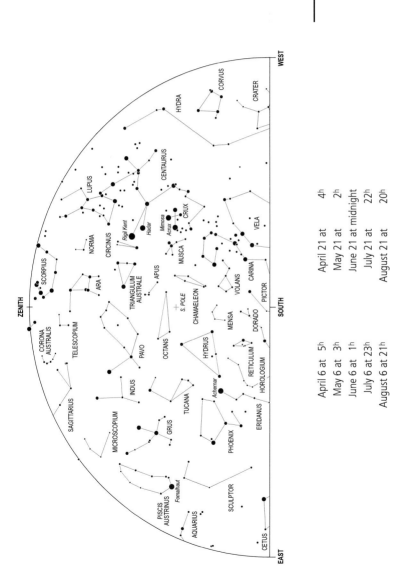

WEST

HYDRA
CORVUS
CRATER
CENTAURUS
LUPUS
Rigil Kent
Hadar
Mimosa
Acrux
CRUX
NORMA
CIRCINUS
VELA
MUSCA
APUS
SCORPIUS
ARA
TRIANGULUM AUSTRALE
CARINA
VOLANS
PICTOR
ZENITH
CORONA AUSTRALIS
TELESCOPIUM
S. POLE
CHAMAELEON
MENSA
DORADO
SOUTH
SAGITTARIUS
PAVO
OCTANS
HYDRUS
RETICULUM
HOROLOGIUM
INDUS
Achernar
MICROSCOPIUM
TUCANA
ERIDANUS
GRUS
PHOENIX
PISCIS AUSTRINUS
Fomalhaut
SCULPTOR
AQUARIUS
CETUS

EAST

April 6 at   5ʰ
May 6 at   3ʰ
June 6 at   1ʰ
July 6 at 23ʰ
August 6 at 21ʰ

April 21 at   4ʰ
May 21 at   2ʰ
June 21 at midnight
July 21 at 22ʰ
August 21 at 20ʰ

8N

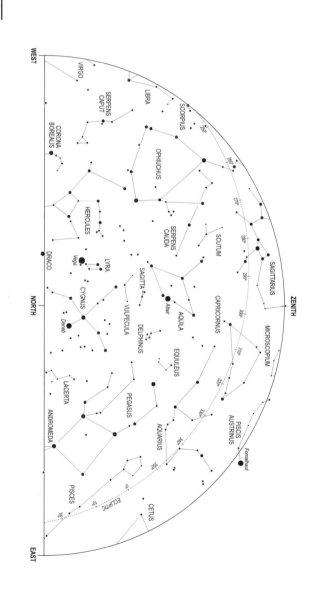

May 6 at 5h
June 6 at 3h
July 6 at 1h
August 6 at 23h
September 6 at 21h

May 21 at 4h
June 21 at 2h
July 21 at midnight
August 21 at 22h
September 21 at 20h

8S

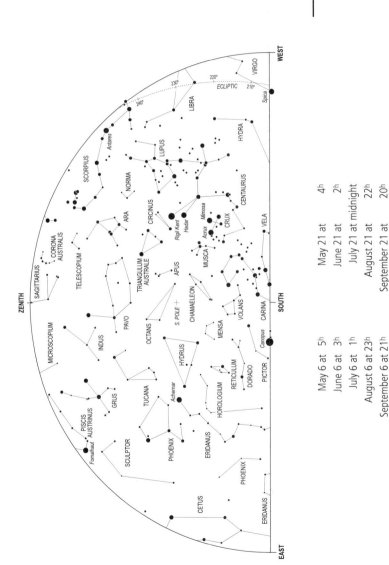

ZENITH

WEST

EAST

SOUTH

WEST
ECLIPTIC
210°
220°
230°
240°
Spica
VIRGO
LIBRA
HYDRA
Antares
SCORPIUS
LUPUS
NORMA
CENTAURUS
ARA
CIRCINUS
Mimosa
CRUX
Rigil Kent
Hadar
Acrux
VELA
CORONA
AUSTRALIS
TELESCOPIUM
TRIANGULUM
AUSTRALE
APUS
MUSCA
SAGITTARIUS
PAVO
S. POLE +
CHAMELEON
VOLANS
CARINA
MICROSCOPIUM
OCTANS
MENSA
Canopus
INDUS
HYDRUS
RETICULUM
PICTOR
GRUS
TUCANA
Achernar
DORADO
HOROLOGIUM
PISCIS
AUSTRINUS
SCULPTOR
PHOENIX
ERIDANUS
Fomalhaut
CETUS
PHOENIX
ERIDANUS

May 21 at   4ʰ
June 21 at   2ʰ
July 21 at midnight
August 21 at   22ʰ
September 21 at 20ʰ

May 6 at   5ʰ
June 6 at   3ʰ
July 6 at   1ʰ
August 6 at   23ʰ
September 6 at 21ʰ

9N

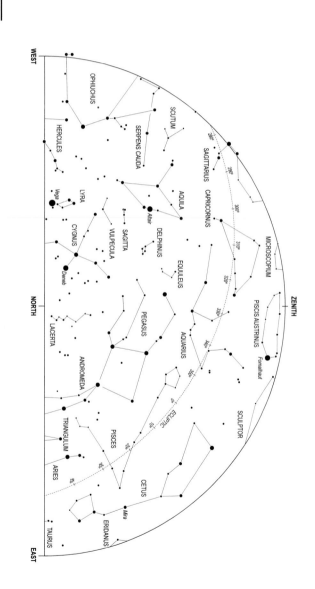

June 6 at 5h
July 6 at 3h
August 6 at 1h
September 6 at 23h
October 6 at 21h

June 21 at 4h
July 21 at 2h
August 21 at midnight
September 21 at 22h
October 21 at 20h

9S

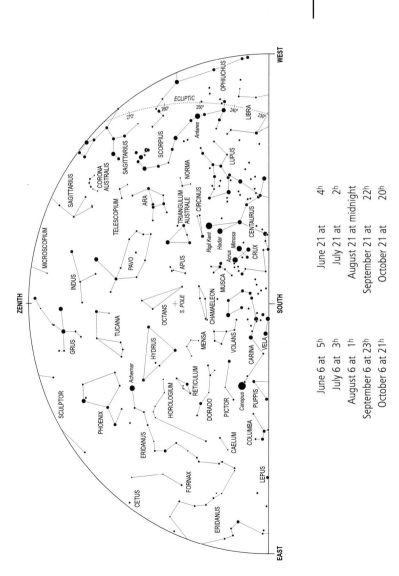

WEST

ECLIPTIC

OPHIUCHUS

270° 260° 250° 240° 230°

Antares

LIBRA

SCORPIUS

SAGITTARIUS

CORONA
AUSTRALIS

LUPUS

NORMA

SAGITTARIUS

TELESCOPIUM

ARA

TRIANGULUM
AUSTRALE

CIRCINUS

CENTAURUS

MICROSCOPIUM

PAVO

APUS

Rigil Kent
Hadar
Acrux Mimosa
CRUX

INDUS

MUSCA

CHAMAELEON

MENSA

OCTANS
+ S. POLE

VOLANS

ZENITH

GRUS

TUCANA

HYDRUS

RETICULUM

CARINA

VELA

SOUTH

SCULPTOR

PHOENIX

Achernar

HOROLOGIUM

DORADO

PICTOR

Canopus

PUPPIS

ERIDANUS

CAELUM

COLUMBA

CETUS

FORNAX

LEPUS

ERIDANUS

EAST

June 21 at   4ʰ
July 21 at   2ʰ
August 21 at midnight
September 21 at 22ʰ
October 21 at 20ʰ

June 6 at 5ʰ
July 6 at 3ʰ
August 6 at 1ʰ
September 6 at 23ʰ
October 6 at 21ʰ

10N

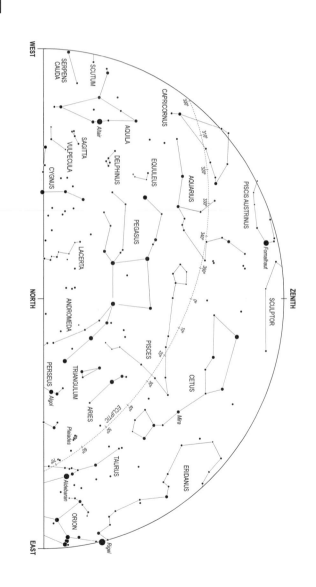

July 6 at 5h
August 6 at 3h
September 6 at 1h
October 6 at 23h
November 6 at 21h

July 21 at 4h
August 21 at 2h
September 21 at midnight
October 21 at 22h
November 21 at 20h

10S

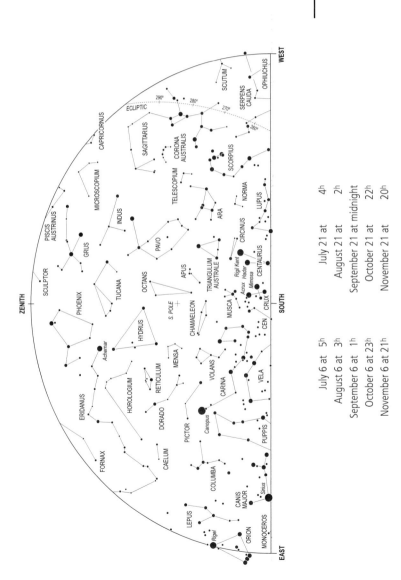

July 21 at 4h
August 21 at 2h
September 21 at midnight
October 21 at 22h
November 21 at 20h

July 6 at 5h
August 6 at 3h
September 6 at 1h
October 6 at 23h
November 6 at 21h

11N

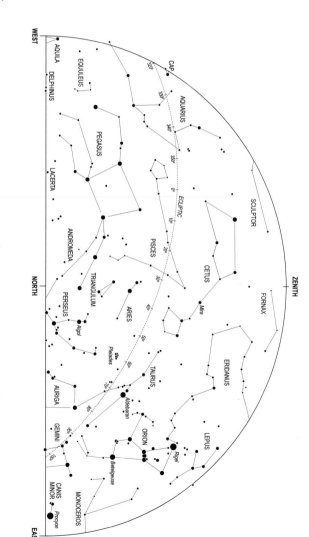

August 6 at 5h
September 6 at 3h
October 6 at 1h
November 6 at 23h
December 6 at 21h

August 21 at 4h
September 21 at 2h
October 21 at midnight
November 21 at 22h
December 21 at 20h

11S

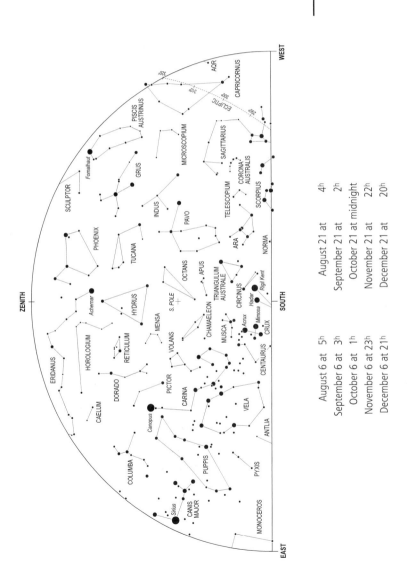

WEST

AQR
CAPRICORNUS
320°
310°
ECLIPTIC
300°
+
290°
PISCIS AUSTRINUS
MICROSCOPIUM
SAGITTARIUS
Fomalhaut
GRUS
SCULPTOR
CORONA AUSTRALIS
INDUS
PAVO
TELESCOPIUM
SCORPIUS
PHOENIX
TUCANA
ARA
NORMA
ZENITH
OCTANS
APUS
Achernar
HYDRUS
S. POLE
TRIANGULUM AUSTRALE
CIRCINUS
Rigil Kent
ERIDANUS
HOROLOGIUM
MENSA
CHAMAELEON
Hadar
SOUTH
RETICULUM
VOLANS
Acrux
Mimosa
MUSCA
CRUX
CAELUM
DORADO
PICTOR
CENTAURUS
CARINA
VELA
Canopus
ANTLIA
COLUMBA
PUPPIS
PYXIS
Sirius
CANIS MAJOR
MONOCEROS

EAST

August 21 at	4ʰ
September 21 at	2ʰ
October 21 at midnight	
November 21 at	22ʰ
December 21 at	20ʰ

August 6 at	5ʰ
September 6 at	3ʰ
October 6 at	1ʰ
November 6 at	23ʰ
December 6 at	21ʰ

12N

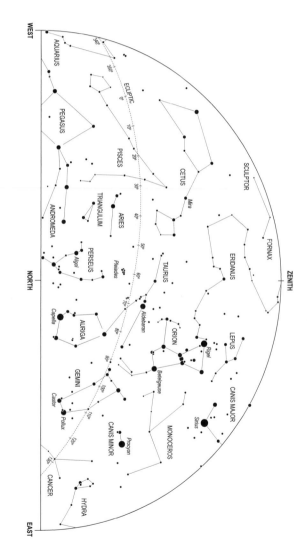

September 6 at 5h
October 6 at 3h
November 6 at 1h
December 6 at 23h
January 6 at 21h

September 21 at 4h
October 21 at 2h
November 21 at midnight
December 21 at 22h
January 21 at 20h

12S

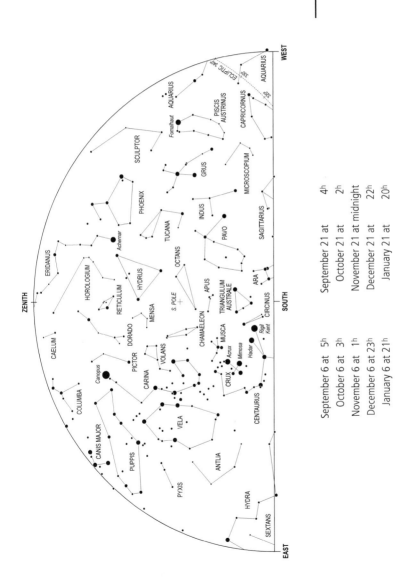

September 21 at 4ʰ
October 21 at 2ʰ
November 21 at midnight
December 21 at 22ʰ
January 21 at 20ʰ

September 6 at 5ʰ
October 6 at 3ʰ
November 6 at 1ʰ
December 6 at 23ʰ
January 6 at 21ʰ

The Planets and the Ecliptic

The paths of the planets about the Sun all lie close to the plane of the ecliptic, which is marked for us in the sky by the apparent path of the Sun among the stars, and is shown on the star charts by a broken line. The Moon and naked-eye planets will always be found close to this line, never departing from it by more than about 7°. Thus the planets are most favourably placed for observation when the ecliptic is well displayed, and this means that it should be as high in the sky as possible. This avoids the difficulty of finding a clear horizon, and also overcomes the problem of atmospheric absorption, which greatly reduces the light of the stars. Thus a star at an altitude of 10° suffers a loss of 60 per cent of its light, which corresponds to a whole magnitude; at an altitude of only 4°, the loss may amount to two magnitudes.

The position of the ecliptic in the sky is therefore of great importance, and since it is tilted at about 23½° to the equator, it is only at certain times of the day or year that it is displayed to the best advantage. It will be realized that the Sun (and therefore the ecliptic) is at its highest in the sky at noon in midsummer, and at its lowest at noon in midwinter. Allowing for the daily motion of the sky, it follows that the ecliptic is highest at midnight in winter, at sunset in the spring, at noon in summer and at sunrise in the autumn. Hence these are the best times to see the planets. Thus, if Venus is an evening object in the western sky after sunset, it will be seen to best advantage if this occurs in the spring, when the ecliptic is high in the sky and slopes down steeply to the horizon. This means that the planet is not only higher in the sky, but will remain for a much longer period above the horizon. For similar reasons, a morning object will be seen at its best on autumn mornings before sunrise, when the ecliptic is high in the east. The outer planets, which can come to opposition (i.e. opposite the Sun), are best seen when opposition occurs in the winter months, when the ecliptic is high in the sky at midnight.

The seasons are reversed in the Southern Hemisphere, spring beginning at the September Equinox, when the Sun crosses the Equator on its way south, summer beginning at the December Solstice, when the

Sun is highest in the southern sky, and so on. Thus, the times when the ecliptic is highest in the sky, and therefore best placed for observing the planets, may be summarized as follows:

	Midnight	Sunrise	Noon	Sunset
Northern latitudes	December	September	June	March
Southern latitudes	June	March	December	September

In addition to the daily rotation of the celestial sphere from east to west, the planets have a motion of their own among the stars. The apparent movement is generally *direct*, i.e. to the east, in the direction of increasing longitude, but for a certain period (which depends on the distance of the planet) this apparent motion is reversed. With the outer planets this *retrograde* motion occurs about the time of opposition. Owing to the different inclination of the orbits of these planets, the actual effect is to cause the apparent path to form a loop, or sometimes an S-shaped curve. The same effect is present in the motion of the inferior planets, Mercury and Venus, but it is not so obvious, since it always occurs at the time of inferior conjunction.

The *inferior planets*, Mercury and Venus, move in smaller orbits than that of the Earth, and so are always seen near the Sun. They are most obvious at the times of greatest angular distance from the Sun (greatest elongation), which may reach 28° for Mercury, and 47° for Venus. They are seen as evening objects in the western sky after sunset (at eastern elongations) or as morning objects in the eastern sky before sunrise (at western elongations). The succession of phenomena, conjunctions and elongations, always follows the same order, but the intervals between them are not equal. Thus, if either planet is moving round the far side of its orbit its motion will be to the east, in the same direction in which the Sun appears to be moving. It therefore takes much longer for the planet to overtake the Sun – that is, to come to superior conjunction – than it does when moving round to inferior conjunction, between Sun and Earth. The intervals given in the table at the top of p. 70 are average values; they remain fairly constant in the case of Venus, which travels in an almost circular orbit. In the case of Mercury, however, conditions vary widely because of the great eccentricity and inclination of the planet's orbit.

		Mercury	**Venus**
Inferior Conjunction	to Elongation West	22 days	72 days
Elongation West	to Superior Conjunction	36 days	220 days
Superior Conjunction	to Elongation East	35 days	220 days
Elongation East	to Inferior Conjunction	22 days	72 days

The greatest brilliancy of Venus always occurs about thirty-six days before or after inferior conjunction. This will be about a month after greatest eastern elongation (as an evening object), or a month before greatest western elongation (as a morning object). No such rule can be given for Mercury, because its distances from the Earth and the Sun can vary over a wide range.

Mercury is not likely to be seen unless a clear horizon is available. It is seldom as much as 10° above the horizon in the twilight sky in northern temperate latitudes, but this figure is often exceeded in the Southern Hemisphere. This favourable condition arises because the maximum elongation of 28° can occur only when the planet is at aphelion (farthest from the Sun), and it then lies well south of the equator. Northern observers must be content with smaller elongations, which may be as little as 18° at perihelion. In general, it may be said that the most favourable times for seeing Mercury as an evening object will be in spring, some days before greatest eastern elongation; in autumn, it may be seen as a morning object some days after greatest western elongation.

Venus is the brightest of the planets and may be seen on occasions in broad daylight. Like Mercury, it is alternately a morning and an evening object, and it will be highest in the sky when it is a morning object in autumn, or an evening object in spring. Venus is to be seen at its best as an evening object in northern latitudes when eastern elongation occurs in June. The planet is then well north of the Sun in the preceding spring months, and is a brilliant object in the evening sky over a long period. In the Southern Hemisphere a November elongation is best. For similar reasons, Venus gives a prolonged display as a morning object in the months following western elongation in October (in northern latitudes) or in June (in the Southern Hemisphere).

The *superior planets*, which travel in orbits larger than that of the Earth, differ from Mercury and Venus in that they can be seen opposite the Sun in the sky. The superior planets are morning objects after conjunction with the Sun, rising earlier each day until they come to opposition. They will then be nearest to the Earth (and therefore at their brightest), and will be on the meridian at midnight, due south in northern latitudes, but due north in the Southern Hemisphere. After opposition they are evening objects, setting earlier each evening until they set in the west with the Sun at the next conjunction. The difference in brightness from one opposition to another is most noticeable in the case of Mars, whose distance from Earth can vary considerably and rapidly. The other superior planets are at such great distances that there is very little change in brightness from one opposition to the next. The effect of altitude is, however, of some importance, for at a December opposition in northern latitudes the planets will be among the stars of Taurus or Gemini, and can then be at an altitude of more than 60° in southern England. At a summer opposition, when the planet is in Sagittarius, it may only rise to about 15° above the southern horizon, and so makes a less impressive appearance. In the Southern Hemisphere the reverse conditions apply, a June opposition being the best, with the planet in Sagittarius at an altitude which can reach 80° above the northern horizon for observers in South Africa.

Mars, whose orbit is appreciably eccentric, comes nearest to the Earth at oppositions at the end of August. It may then be brighter even than Jupiter, but rather low in the sky in Aquarius for northern observers, though very well placed for those in southern latitudes. These favourable oppositions occur every fifteen or seventeen years (e.g. in 1988, 2003, 2018), but in the Northern Hemisphere the planet is probably better seen at oppositions in the autumn or winter months, when it is higher in the sky. Oppositions of Mars occur at an average interval of 780 days, and during this time the planet makes a complete circuit of the sky.

Jupiter is always a bright planet, and comes to opposition a month later each year, having moved, roughly speaking, from one Zodiacal constellation to the next.

Saturn moves much more slowly than Jupiter, and may remain in the same constellation for several years. The brightness of Saturn depends on the aspects of its rings, as well as on the distance from the Earth and Sun. The Earth passed through the plane of Saturn's rings in 1995 and 1996, when they appeared edge-on; we saw them at maximum opening, and Saturn at its brightest, in late 2002. The rings will next appear edge-on in 2009.

Uranus and *Neptune* are both visible with binoculars or a small telescope, but you will need a finder chart to help locate them, while *Pluto* is hardly likely to attract the attention of observers without adequate telescopes.

Phases of the Moon in 2004

Full Moon				Last Quarter				New Moon				First Quarter		
	d	h	m		d	h	m		d	h	m		d	h	m
Jan	7	15	40	Jan	15	04	46	Jan	21	21	05	Jan	29	06	03
Feb	6	08	47	Feb	13	13	39	Feb	20	09	18	Feb	28	03	24
Mar	6	23	14	Mar	13	21	01	Mar	20	22	41	Mar	28	23	48
Apr	5	11	03	Apr	12	03	46	Apr	19	13	21	Apr	27	17	32
May	4	20	33	May	11	11	04	May	19	04	52	May	27	07	57
June	3	04	20	June	9	20	02	June	17	20	27	June	25	19	08
July	2	11	09	July	9	07	34	July	17	11	24	July	25	03	37
July	31	18	05	Aug	7	22	01	Aug	16	01	24	Aug	23	10	12
Aug	30	02	22	Sept	6	15	10	Sept	14	14	29	Sept	21	15	54
Sept	28	13	09	Oct	6	10	12	Oct	14	02	48	Oct	20	21	59
Oct	28	03	07	Nov	5	05	53	Nov	12	14	27	Nov	19	05	50
Nov	26	20	07	Dec	5	00	53	Dec	12	01	29	Dec	18	16	40
Dec	26	15	06												

All times are GMT

Longitudes of the Sun, Moon and Planets in 2004

Date		Sun °	Moon °	Venus °	Mars °	Jupiter °	Saturn °
January	6	285	87	319	12	169	99
	21	300	288	338	22	168	98
February	6	317	132	357	32	167	97
	21	332	340	14	41	166	97
March	6	346	154	30	50	164	96
	21	1	1	46	60	162	96
April	6	17	204	62	70	160	97
	21	31	47	75	80	159	98
May	6	46	242	84	89	159	99
	21	60	80	86	99	159	101
June	6	76	296	79	109	160	103
	21	90	124	71	118	162	105
July	6	104	333	70	128	164	106
	21	119	158	77	137	167	108
August	6	134	22	89	147	170	110
	21	148	207	103	157	173	112
September	6	164	67	119	167	176	114
	21	178	260	135	177	179	115
October	6	193	98	153	186	182	116
	21	208	299	170	196	186	117
November	6	224	142	190	207	189	117
	21	239	351	208	217	192	117
December	6	254	175	227	227	194	117
	21	269	27	245	237	196	116

Longitude of *Uranus* 334°
Neptune 314°

Moon: Longitude of ascending node
Jan 1: 48°   Dec 31: 28°

Mercury moves so quickly among the stars that it is not possible to indicate its position on the star charts at convenient intervals. The monthly notes must be consulted for the best times at which the planet may be seen.

The positions of the other planets are given in the table on p. 74. This gives the apparent longitudes on dates which correspond to those of the star charts, and the position of the planet may at once be found near the ecliptic at the given longitude.

EXAMPLES

In the Southern Hemisphere two planets are seen close together in the north-western sky in the early evening, in late May. Identify them.

The Southern Star Chart 3N shows the north-western sky on May 21 at 18h and shows longitudes 78° to 220° along the ecliptic. Reference to the table on p. 74 gives the longitude of Venus as 86°, though the star chart shows that it is only a few degrees above the horizon and is unlikely to be seen. The table also gives the longitude of Mars as 99° and that of Saturn as 101°. Thus these two planets are to be found close together in the north-western sky, and the brighter one is Saturn.

The positions of the Sun and Moon can be plotted on the star maps in the same manner as for the planets. The average daily motion of the Sun is 1°, and of the Moon 13°. For the Moon an indication of its position relative to the ecliptic may be obtained from a consideration of its longitude relative to that of the ascending node. The latter changes only slowly during the year, as will be seen from the values given on p. 74. Let us denote by d the difference in longitude between the Moon and its ascending node. Then if $d = 0°$, 180° or 360°, the Moon is on the ecliptic. If $d = 90°$ the Moon is 5° north of the ecliptic, and if $d = 270°$ the Moon is 5° south of the ecliptic.

On October 21 the Moon's longitude is given in the table on p.74 as 299° and the longitude of the node is found by interpolation to be about 32°. Thus $d = 267°$, and the Moon is about 5° south of the ecliptic. Its position may be plotted on northern star charts 7S, 8S, 9S and 10S, and on southern star charts 7N, 8N and 9N.

Some Events in 2004

ECLIPSES

There will be four eclipses, two of the Sun and two of the Moon.

April 19:	partial eclipse of the Sun – southern Africa
May 4:	total eclipse of the Moon – Australasia, Asia, Europe, Africa, South America
October 14:	partial eclipse of the Sun – Asia
October 28:	total eclipse of the Moon – Asia, Africa, Europe, the Americas

THE PLANETS

Mercury may be seen more easily from northern latitudes in the evenings about the time of greatest eastern elongation (March 29) and in the mornings about the time of greatest western elongation (September 9). In the Southern Hemisphere the corresponding most favourable dates are around May 14 (mornings) and July 27 (evenings).

Venus is visible in the evenings from the beginning of the year until May and in the mornings from July onwards.

Mars does not come to opposition in 2004.

Jupiter is at opposition on March 4 in Leo.

Saturn does not come to opposition in 2004.

Uranus is at opposition on August 27 in Aquarius.

Neptune is at opposition on August 6 in Capricornus.

Pluto is at opposition on June 11 in Serpens.

Monthly Notes 2004

January

New Moon: January 21   *Full Moon:* January 7

EARTH is at perihelion (nearest to the Sun) on January 4 at a distance of 147 million kilometres (91 million miles).

MERCURY attains its greatest western elongation (24°) on January 17. Observers in southern and equatorial latitudes will be able to see the planet after the first few days of the month, low in the south-eastern sky before dawn. During this period its magnitude brightens from +0.9 to −0.2. However, for observers in the latitudes of the British Isles, Mercury will be a very difficult morning object at a very low altitude (around 5° at the beginning of morning civil twilight, and then only until shortly after the middle of the month).

VENUS is a brilliant object in the south-western sky after sunset, magnitude −4.0. Observers in the latitudes of the British Isles will note that the period of observation increases from two to three hours during the month as the planet moves northwards in declination and also increases its angular distance from the Sun.

MARS is an evening object, and will be seen in the south in the early evening twilight, setting around midnight over the western horizon for observers in the latitudes of the British Isles, but several hours earlier in temperate southern latitudes. The planet is fading quite noticeably as its distance from the Earth increases, the magnitude fading from +0.2 to +0.7. There will be no opposition of Mars in 2004 and throughout the year it moves quite rapidly round the sky as will be seen from Figures 8 and 26, given with the notes for April and November respectively.

JUPITER, magnitude −2.3, is visible well before midnight as a brilliant object in the night sky, and is west of the meridian well before dawn. Jupiter is in Leo, reaching its first stationary point on January 3, and then moving slowly retrograde. Figure 3, given with the notes for February, shows its path among the stars throughout the year.

SATURN was at opposition on December 31 last, and is therefore visible for the greater part of the hours of darkness, being seen in the eastern sky as soon as it is dark. It is retrograding very slowly in Gemini. Saturn's magnitude is −0.4. Figure 1 shows its path among the stars throughout the year. There is no opposition of Saturn in 2004.

Saturn and its Satellites. Saturn is now well placed for observation. With its magnificent system of rings (Figure 2), it is often regarded as the most beautiful object in the entire sky. Although the rings are now just past their best (having been at their maximum opening in late 2002), they are still superbly displayed. But do not forget Saturn's satellites, which are fascinating worlds in their own right. Indeed, as the text for this *Yearbook* was being put together, a team of astronomers announced the discovery of a new satellite orbiting Saturn, bringing its moon total up to thirty-one. This latest addition to Saturn's retinue is a small rocky body only 8 kilometres (5 miles) across. It follows a very elliptical retrograde orbit that carries it far away from Saturn itself.

Saturn's thirty-first satellite was first discovered on February 5, 2003 by Scott Shepard and David Jewitt of the University of Hawaii and Jan Kleyna of Cambridge University. Follow-up observations were made with the telescopes on Mauna Kea, using orbital calculations provided by Brian Marsden of the Harvard Smithsonian Center for Astrophysics.

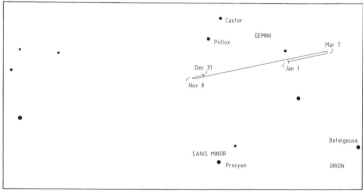

Figure 1. The path of Saturn against the stars of Gemini and Cancer during 2004.

Figure 2. Image of Saturn acquired by the Hubble Space Telescope's infrared camera. It provides detailed information on the clouds and hazes in Saturn's atmosphere. Most of the Northern Hemisphere that is visible above the rings is relatively clear. The dark region around the South Pole indicates a big hole in the main cloud layer. (Image courtesy Erich Karkoschka (University of Arizona), and NASA/STScI.)

The discovery of this new moon was formally announced on April 8, 2003. (Incidentally, teams led by the same astronomers were also responsible for the discovery of twenty new moons around Jupiter in the first four months of 2003, bringing its total to sixty!)

Only eight of Saturn's moons are generally visible with small or moderate telescopes, and these are:

Name	Mean distance from Saturn km	Mean Orbital Period days	Diameter km	Mean Visual Magnitude
Mimas	185,520	0.942	421 × 395 × 385	12.9
Enceladus	238,020	1.370	512 × 495 × 488	11.7
Tethys	294, 660	1.888	1,046	10.2
Dione	377,400	2.737	1,120	10.4
Rhea	527,040	4.518	1,528	9.7
Titan	1,221,850	15.945	5,150	8.3
Hyperion	1,481,100	21.277	410 × 260 × 220	14.2
Iapetus	3,561,300	79.331	1,460	10.2–11.9

Hyperion is not an easy object (it is also highly irregular in shape). Another satellite, Phoebe, which orbits at a mean distance of almost

13 million kilometres from Saturn, at magnitude 16.5, is beyond the range of average amateur telescopes. Apart from Titan, the satellites are icy and cratered; Iapetus has one hemisphere which is bright and one which is dark, and displays marked variations in brightness; it is always at its brightest when to the west of Saturn. The nature of the dark coating is not definitely known.

At present, the Cassini–Huygens spacecraft, launched in October 1997, is drawing ever nearer to Saturn. It will arrive on July 1, 2004, firing its engine to slow its speed so that it can be captured into a closed orbit around Saturn. If all goes well, Cassini–Huygens will fly by Saturn's largest moon, Titan, on October 26. Then, on December 24, it will release the Huygens probe, which will free-fall to Titan, entering the dense atmosphere of the satellite on January 14, 2005 and sending back data for a brief period. Whether it will land on a solid surface, or plunge into a chemical ocean, remains to be seen. Thereafter, the Cassini orbiter will continue moving around Saturn for four years, sending back data of all kinds. These notes are being written in April 2003, so we can only hope that the schedule will be successfully followed. For a detailed review of the Cassini–Huygens mission, and of its arrival at Saturn, see the article by David M. Harland elsewhere in this *Yearbook*.

This Month's Anniversary. John Pringle Nichol, the eldest son of a gentleman farmer, was born at Huntly-Hill, near Brechin, Scotland, on January 13, 1804, and educated at Brechin Grammar School, before obtaining the highest honours in mathematics and physics at Kings College, Aberdeen University. After a brief spell as a church minister, he decided on a career in education, becoming headmaster of Hawick Grammar School and Cupar Academy, then Rector of Montrose Academy. In 1836, he was appointed Regius Professor of Astronomy in the University of Glasgow, as well as Director of the Observatory there, and became well known as an astronomer.

Nichol was, by all accounts, an enthusiastic and popular lecturer and an inspiring teacher. He extended his work beyond the University, presenting public lectures about science to crowded meetings in the city. As an author, he established a lasting reputation; his book *The Architecture of the Heavens* appeared in 1838, and ran to seven editions. Soon after, *The Solar System* appeared, later republished as *The Planetary System*. *The System of the World* followed in 1846, then *The Stellar Universe* and, in 1855, *The Planet Neptune*, a book based on one of the author's most popular lectures.

Although Nichol was principally a lecturer and popularizer, rather than a researcher, he certainly had the gifts needed for a first-class practical astronomer. He was evidently a strong supporter of Laplace's nebular hypothesis for the origin of the Solar System (in which the planets condense from different zones within a nebula). Nichol was twice married, and died in Rothesay on September 19, 1859, at the age of fifty-five.

February

MERCURY is not visible to observers in northern temperate latitudes, but for observers further south continues to be visible low in the south-eastern sky before dawn for the first half of the month. During this period its magnitude brightens slowly from −0.2 to −0.6.

VENUS, magnitude −4.1, continues to be visible as a brilliant evening object. It is visible in the south-western sky for several hours after sunset.

MARS continues to be visible as a evening object in the south-western sky, its magnitude fading during the month from +0.7 to +1.1. As it moves northwards in declination, it has to cover a longer arc in its daily apparent motion across the sky, and as a result it continues to set at about the same time (around midnight for observers in the latitudes of the British Isles) during February and March. Mars is in the constellation of Aries. The reddish tinge of Mars is an aid to its identification.

JUPITER, magnitude −2.5, comes to opposition early next month and therefore will be seen rising in the east shortly after sunset, and remaining visible right through until dawn. Jupiter is retrograding slowly in the constellation of Leo, as will be seen from the accompanying diagram (Figure 3) indicating its path against the stars during the year.

SATURN continues to be visible in the south-western quadrant of the sky in the evenings, and for observers in northern temperate latitudes it will still be seen for several hours after midnight, even at the end of the month. It is retrograding slowly in Gemini. Saturn's magnitude is −0.2.

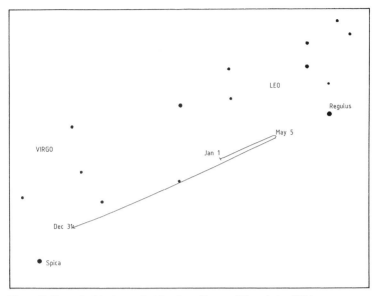

Figure 3. The path of Jupiter against the stars of Leo and Virgo during 2004.

Aries and Triangulum. Throughout this month, Mars lies in the constellation of Aries, the Ram. Aries is still often regarded as the first constellation of the Zodiac, even though the so-called 'first point of Aries' – the point on the celestial equator from which right ascension (the celestial equivalent of longitude on Earth) is measured – has now been shifted westwards by precession into the adjacent constellation of Pisces.

In mythology, Aries represents the ram with the golden fleece, sent by the god Mercury to rescue the two children of the King of Thebes from the clutches of their evil stepmother who was planning to kill them. The ram which, rather remarkably, could fly, carried out its mission but, unfortunately, the girl (Helle) fell from the ram to her death in that region of the sea now known as the Hellespont, although the boy arrived safely. When the ram died, its fleece was hung in a sacred grove, from where it was later removed by Jason and the Argonauts.

Though Mars is now fading, it is still a magnitude brighter than Hamal or Alpha Arietis, the brightest star in the Ram. In fact, Aries is not very conspicuous, but it is fairly easy to identify; the three main stars are Alpha (magnitude 2.0), Beta or Sheratan (2.6) and Gamma or

Mesartim (3.9). Alpha has a K-type spectrum, and so is decidedly orange-red, though its colour cannot compare with the reddish hue of Mars. The most interesting object in the constellation is Gamma, which is a particularly wide and easy double; the components are both white, exactly equal at magnitude 4.8, and the separation is 7.5 seconds of arc.

There are a couple of Mira-type variable stars in Aries, both of which can be as bright as the seventh magnitude at maximum, but visible only in a moderate-to-large telescope at minimum: R Arietis has a range between 7.4 and 13.7, and a mean period of 187 days, while U Arietis has range from 7.2 to 15.2, and a mean period of 371 days.

Triangulum, lying between Aries and Andromeda (Figure 4), is one of the few constellations to resemble the object after which it is named. Alpha or Rasalmothallah (magnitude 3.4), Beta (3.0) and Gamma (4.0) do indeed form a sharp-pointed triangle with Alpha at the apex. The spiral galaxy M33 lies not far from Alpha, and is on the fringe of naked-eye visibility; it is much the brightest of the spirals apart from M31, the great spiral in Andromeda and, just like M31, M33 is a member of the Local Group of galaxies to which our own Milky Way belongs. Current

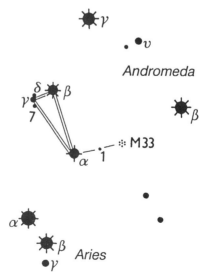

Figure 4. The principal stars of Aries and Triangulum, showing the location of the spiral galaxy M33.

estimates put the distance of M33 at just under three million light years. With ×7 binoculars, M33 is easy to find; look past the faint star 1 Trianguli and, at about an equal distance beyond it in the direction of Beta Andromedae, M33 will show up as a hazy patch. Telescopically it is strangely elusive because of its relatively low surface brightness. Very keen-sighted people claim to have glimpsed it with the naked eye under very clear, dark conditions.

This Month's Anniversaries. This month marks the anniversaries of the deaths of two notable individuals. Immanuel Kant, the famous German philosopher (and considered by some to be the most influential thinker of modern times), was born in Königsberg (now Kaliningrad, Russia) on April 22, 1724. Having first studied mainly classics, Kant took up physics and mathematics at the University of Königsberg and, after obtaining his doctorate, taught science and mathematics at the University for some fifteen years. Later, he expanded his areas of interest to cover all branches of philosophy, becoming Professor of Logic and Metaphysics at the University in 1770.

Astronomically, Kant is remembered for proposing a simple theory for the origin of the Solar System in his 1755 work *Universal Natural History and Theories of the Heavens*. Kant described how planets form in what is today known as a protoplanetary disc, explaining that within such a disc, 'the attraction of the elementary matters for each other' drew particles together to form larger and larger objects, eventually building planet-sized bodies. Kant's ideas were later taken up by the French mathematician Pierre-Simon de Laplace in his so-called nebular theory of planetary formation. Confirmation of part of the Kant–Laplace hypothesis came in 1983 with the discovery of a rotating disc of dust around the young star Beta Pictoris, and their ideas form the basis of all modern work on planet formation. Kant died on February 12, 1804.

The French astronomer Henri Perrotin was born at St Loup on December 19, 1845, and worked at Toulouse before becoming the first Director of the Nice Observatory. He discovered several asteroids, but is remembered chiefly for his planetary work. In 1877 the Italian astronomer Giovanni Schiaparelli had first reported an alleged network of 'canals' of Mars, which for some years were not seen by anybody else; Schiaparelli had called them *canali* (channels), but this was invariably translated as canals. In 1886, while working with his colleague

Louis Thollon on the 38-cm (15-inch) refractor at Nice, Perrotin reported seeing many 'canals' on Mars in positions agreeing with the map produced by Schiaparelli in 1882. Subsequently they were widely reported, though we now know that they do not exist. It seems that once Schiaparelli had shown observers how to *see* Mars, it then became impossible to see it any other way. The expectation of the observers was so high that they were easily deceived.

Perrotin sketched more canals in 1888, using the new 76-cm (30-inch) refractor at Nice, then the largest in Europe. He also reported dramatic changes (since 1886) in an area on Mars named 'Libya', believing it to have been flooded by a neighbouring sea. Subsequently, Perrotin's observations were challenged by astronomers using even more powerful instruments, such as the 91-cm (36-inch) Clark refractor at the Lick Observatory in California. Perrotin died at Nice on February 29, 1904.

March

New Moon: March 20 *Full Moon:* March 6

Equinox: March 20

Summer Time in the United Kingdom commences on March 28.

MERCURY reaches superior conjunction on March 4 and is therefore unsuitably placed for observation during the first part of the month. Shortly after the middle of the month it becomes an evening object for observers in equatorial and northern latitudes. For observers in northern temperate latitudes this will be the most favourable evening apparition of the year. Figure 5 shows, for observers in latitude 52°N, the changes in azimuth (true bearing from the north through east, south and west) and altitude of Mercury on successive evenings when the Sun is 6° below the horizon. This condition is known as the end of evening civil twilight and in this latitude and at this time of year occurs about thirty-five minutes after sunset. The changes in the brightness of the planet are indicated by the relative sizes of the circles marking Mercury's position at five-day intervals. It will be noticed that Mercury is at its brightest before it reaches greatest eastern elongation (19°) on March 29. At the beginning of the period of visibility its magnitude is −1.2, but this will have faded to +0.4 by the end of the month.

VENUS, magnitude −4.3, continues to be a brilliant evening object in the west-north-western sky. It reaches its greatest eastern elongation (46°) on March 29.

MARS, magnitude +1.3, remains an evening object in the western sky, setting slightly north of west around midnight. During the month Mars moves eastwards from Aries into Taurus, passing south of the Pleiades.

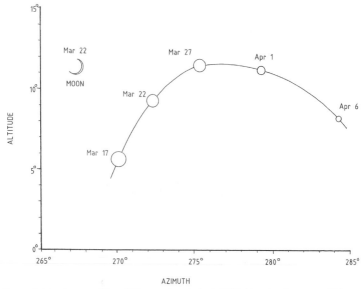

Figure 5. Evening apparition of Mercury, from latitude 52°N. (Angular diameters of Mercury and the crescent Moon not to scale.)

JUPITER reaches opposition on March 4 and is therefore visible as a brilliant object in Leo throughout the hours of darkness. Its magnitude is −2.5.

SATURN, magnitude 0.0, is still to be seen as an evening object in the western sky. It reaches its second stationary point on March 7, and then resumes its direct motion, eastwards, in Gemini.

MESSENGER to Mercury. Though Mercury is well placed for observers in equatorial and more northerly latitudes this month, ordinary telescopes will show virtually no surface detail. Before the Space Age, the best map was due to E.M. Antoniadi, who used the 83-cm (33-inch) refractor at the Observatory of Paris-Meudon. He was an expert observer, but even so his map of Mercury proved to be very inaccurate. There has so far been only one space mission, Mariner 10, which made three active passes of the planet on March 29, 1974, September 21, 1974 and March 16, 1975. Excellent images were obtained (Figure 6), showing the mountains, numerous craters and other features; the most

Figure 6. Mariner 10 mosaic of Mercury taken on the approach to the first encounter in March 1974. It comprises eighteen pictures taken at 42-second intervals, six hours before closest approach. The lower two-thirds of the visible portion is the southern hermisphere of Mercury. The smallest features visible are about 2km across. (Image courtesy NASA/NSSDC.)

imposing structure, the Caloris Basin (a multi-ringed impact basin), is 1,300 kilometres (800 miles) in diameter. However, Mariner 10's coverage of the surface was incomplete, the same areas of the planet's surface being sunlit during each of the three Mariner passes.

In many ways, Mercury is the least explored of the terrestrial planets. Its density is the highest of any planet, suggesting that the planet is composed of about 70 per cent iron, probably concentrated in the core, and it possesses a slight, but measurable, global magnetic field. Mercury's ancient surface records impact events from the earliest stages of planetary formation. Its exotic atmosphere (mainly hydrogen and helium temporarily captured from the solar wind) is the thinnest among those of all the terrestrial planets. And temperatures on this planet closest to the Sun vary from second highest in the Solar System (430°C) at the 'hot spots' on the equator, to among the coldest (-180°C), in the permanently shadowed poles.

All being well, a new US space mission, MESSENGER (MErcury Surface Space ENvironment GEochemistry and Ranging) is due for launch on a Delta II rocket this month (between March 10 and 29), but if the launch is delayed for any reason, there is another 'launch window' in May (between May 12 and 23). MESSENGER will follow a complex path, and will not finally reach Mercury for five years; it makes two fly-bys of Venus (on June 24, 2004 and March 16, 2006) and two fly-bys of Mercury (on July 21, 2007 and April 11, 2008) before entering orbit around Mercury on April 6, 2009. During its two fly-bys of Mercury, MESSENGER will map nearly the entire planet in colour, imaging most of the part unseen by Mariner 10, and take reconnaissance measurements of surface, atmosphere and magnetosphere composition. These fly-by results will be invaluable in planning the orbital mission.

Once in orbit around the planet, MESSENGER is scheduled to carry out comprehensive measurements for one Earth year. MESSENGER's orbit about Mercury is to be highly elliptical, 200 kilometres (120 miles) above the surface at its lowest point and more than 15,000 kilometres (9,000 miles) at its highest. The plane of the orbit is to be inclined 80° to Mercury's rotation axis, and the low point in the orbit is reached at latitude of 60°N. This low northern hemisphere altitude allows for detailed measurements of the geology and composition of the giant Caloris impact basin. Data collection concludes in April 2010. Hostile and lifeless though it undoubtedly is, Mercury is a very interesting world and, as yet, our knowledge of it is very far from complete.

At its brightest, Mercury is of magnitude −1.9, brighter than any star, but of course it is never seen against a really dark background, even during the fleeting moments of a total solar eclipse.

The Old Pole Star. Most people can identify the present north pole star, Polaris (magnitude 2.0) in Ursa Minor (the Little Bear), but how many people can find the old pole star? At the time when the Egyptian Pyramids were being built, the north celestial pole lay near Thuban, in the constellation of Draco (the Dragon). In fact, the Earth's north (and south) celestial pole takes about 25,800 years to describe a complete circle on the celestial sphere. The reason is that the axis of rotation of the spinning Earth is precessing slowly (like the wobbling motion of a gyroscope or child's spinning top), due to gravitational perturbations by the Sun, Moon and planets.

Though Thuban has been given the Greek letter Alpha, it is only the eighth brightest star in the Dragon; its magnitude is 3.7, whereas the leader of the constellation, Eltamin or Gamma Draconis, located in

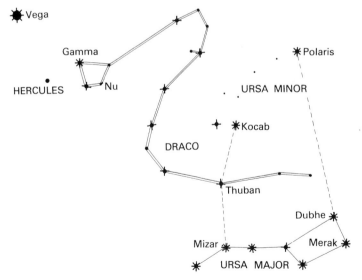

Figure 7. Thuban was the closest naked eye star to the north celestial pole 5,000 years ago. It lies roughly midway between Kocab in Ursa Minor and Mizar in the 'handle' of the Plough.

the Dragon's head, is of magnitude 2.2. However, Thuban is a fairly considerable star; it is 130 times as luminous as the Sun, and 230 light years away. It is pure white, and of spectral type A0, so that its surface is much hotter than that of the Sun.

Identifying Thuban is easy enough, as shown in Figure 7. Find Mizar in the Great Bear, and then Kocab or Beta Ursae Minoris in the Little Bear; Thuban lies more or less midway between them, and there are no other bright stars close to it.

The closest approach of Polaris to the north celestial pole will be in March 2100. Its declination will then be +89°, 32 minutes, 51 seconds. Polaris is actually a Cepheid variable star with a very small amplitude. It has a ninth magnitude companion, with a separation of eighteen arcseconds.

There is no bright south pole star now; the pole lies near the obscure Sigma Octantis, magnitude 5.5. In Pyramid times there was a brighter south pole star, Alpha Hydri (magnitude 3.0), but the star was several degrees from the actual pole. For southern observers, things will be decidedly better by AD 5000, when the pole will be fairly close to Miaplacidus or Beta Carinae, magnitude 1.7.

April

MERCURY is visible as an evening object, very low in the western sky at the end of evening civil twilight, for observers in equatorial and northern temperate latitudes, though only for the first few days of the month. Its magnitude fades rapidly from +0.6 to +1.3 during this time. Mercury passes through inferior conjunction on April 17 and remains too close to the Sun for observation for the rest of the month.

VENUS continues to be visible as a brilliant evening object, magnitude −4.5, completely dominating the western sky for several hours after sunset. For observers in the British Isles Venus will not be setting until after 23h.

MARS, magnitude +1.5, is still visible as an evening object in Taurus, passing north of the Hyades early in the month. It will then be noted that Aldebaran is about 0.6 magnitudes brighter than the planet. Figure 8 shows the path of Mars among the stars during the first half of the year.

JUPITER continues to be visible as a brilliant object in the evening sky, magnitude −2.3. The four Galilean satellites are readily observable in a small telescope or even a good pair of binoculars provided that they are held rigidly.

SATURN is still visible as an evening object in the western sky. By the end of the month it is too low in the west to be seen after midnight. The planet is in the constellation of Gemini, magnitude +0.1. The rings of Saturn present a beautiful spectacle to the observer with a small telescope. They were open to their maximum extent in late 2002, but even now the rings extend only slightly less than the polar diameter of the planet.

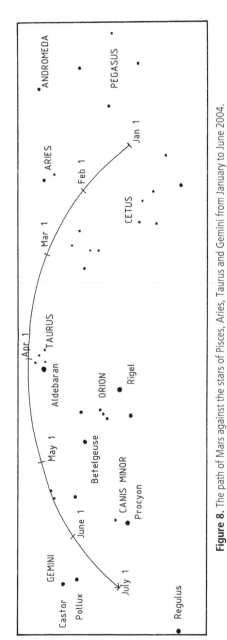

Figure 8. The path of Mars against the stars of Pisces, Aries, Taurus and Gemini from January to June 2004.

Mars and Aldebaran. Mars is well known as the Red Planet, and its colour is very pronounced – which is why, of course, it was named after the God of War. Few stars are of comparable hue, although the red supergiant Antares in Scorpius (the Scorpion) is a candidate; its name means 'Rival of Ares' (Mars) and it is of spectral type M. This month Mars is not far from Aldebaran in Taurus, and it is interesting to compare the two. Aldebaran is not a supergiant star; it is a normal giant of spectral type K, and is orange-red in colour rather than red.

Can You Find the Centaur? Centaurus, the Centaur, is one of the most magnificent constellations in the sky. It contains the Southern Pointers, Alpha and Beta Centauri, which show the way to the Southern Cross, and here too are many important objects: there is the unusual radio galaxy NGC 5128, better known as Centaurus A, some lovely open star clusters, including that around the star Lambda Centauri, and Omega Centauri, generally considered the finest globular cluster in the entire sky. Unfortunately for British and North American observers, Centaurus lies well into the Southern Hemisphere of the sky.

To see whether a star rises from any particular location, you need to know its declination, i.e. its angular distance north or south of the celestial equator. For example, in round figures the declination of Alpha Centauri is −61°. Subtract 61 from 90 and the answer is 29. Therefore Alpha Centauri will never rise from a location on the Earth's surface north of latitude 29°N, and will never set from a point south of latitude 29°S. Between these two latitudes, the star will rise and set. In practice these figures must be slightly modified to allow for atmospheric refraction, but they are accurate enough for most purposes.

The declinations (in round figures) of some of the stars shown in Figure 9 (and the magnificent globular cluster Omega Centauri) are:

Theta Centauri, −36°
Eta Centauri, −42°
Omega Centauri, −47°
Gamma Centauri, −49°
Beta Centauri, −60°
Alpha Centauri, −61°

So can you see any part of the Centaur from Britain? Go to Lizard Point, and you may glimpse Theta (Haratan), magnitude 2.1. Look for

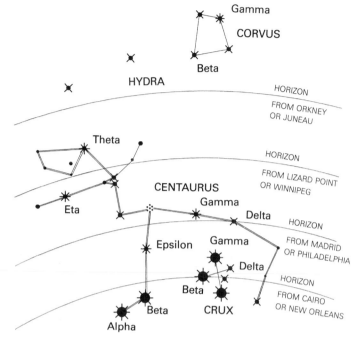

Figure 9. From most of Britain and the northern parts of North America, the pattern of Centaurus never rises high above the horizon, but it is a magnificent constellation when properly seen from locations further south.

it around midnight in early April and, given a really clear southern horizon, you may be lucky.

Of course, the Southern Pointers remain far below the horizon from Britain. Alpha is the closest star beyond the Sun, at just over four-and-a-quarter light years; Beta is a very luminous giant, lying at a distance of about 460 light years. Beta is called either Agena or Hadar; rather strangely, Alpha has never been given a universally accepted proper name – Toliman or Rigel Kent are sometimes used, but astronomers in general refer to it simply as Alpha Centauri.

May

New Moon: May 19   *Full Moon:* May 4

MERCURY, although it reaches greatest western elongation (26°) on May 14, is not suitably placed for observation by those in higher northern latitudes. For observers further south this will be the most favourable morning apparition of the year. Figure 10 shows, for observers in latitude 35°S, the changes in azimuth (true bearing from the north through east, south and west) and altitude of Mercury on successive evenings when the Sun is 6° below the horizon. This condition is known as the beginning of morning civil twilight and in this latitude and at this time of year occurs about thirty minutes before sunrise. The changes in the brightness of the planet are indicated by the relative sizes of the circles marking Mercury's position at five-day intervals. It will be noticed that Mercury is at its brightest after it reaches greatest western elongation.

VENUS is a brilliant object, continuing to dominate the western sky in the evenings, magnitude −4.5. It attains its greatest brilliancy on May 2. The period of time available for observation is decreasing, particularly for observers in the latitudes of the British Isles where it sets about two hours earlier at the end of the month than at the beginning. The telescopic appearance of the planet changes quite markedly during the month. At the beginning of May its apparent diameter is 36 arcseconds and it is 29 per cent illuminated: at the end its apparent diameter has increased to 56 arcseconds while it is then only 3 per cent illuminated, a beautifully thin sliver of light. Venus is occulted by the thin crescent Moon, only two days old, on May 21. The occultation is visible from the British Isles, but only telescopically, since it occurs in the middle of the day.

MARS, magnitude +1.7, continues to be visible as an evening object in the western sky, moving from Taurus into Gemini early in the month.

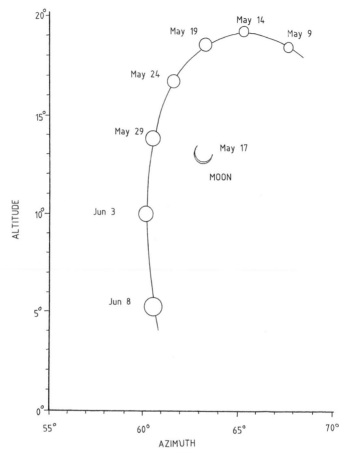

Figure 10. Morning apparition of Mercury, from latitude 35°S. (Angular diameters of Mercury and the crescent Moon not to scale.)

It is moving steadily towards Saturn, passing 1.6° north of it on May 25, Saturn then being 1.5 magnitudes brighter than Mars.

JUPITER, magnitude −2.1, is still visible as a brilliant object in the evenings in the western sky. On May 5, Jupiter reaches its second stationary point and resumes its direct motion, moving slowly eastwards in Leo. By the end of the month it is not visible for long after midnight.

SATURN, magnitude +0.2, is now coming towards the end of its evening apparition, and can only be seen for a short time after dusk, low in the western sky.

This Month's Total Eclipse of the Moon. There are two total eclipses of the Moon this year, after which we must wait until March 3, 2007. Whereas an eclipse of the Sun is seen over only a small portion of the day side of Earth, an eclipse of the Moon is visible from any location on the night side of Earth from where the Moon is above the horizon; this means that for any particular point on Earth, eclipses of the Moon are more common than those of the Sun.

Eclipses of the Moon are caused by the Moon's entry into the cone of shadow cast by the Earth. The accompanying diagram (Figure 11) shows how lunar eclipses occur; clearly they can only occur at Full Moon. To either side of the main central cone of the Earth's shadow (the umbra), there is an area of partial shadow (the penumbra), and obviously the Moon must pass through the penumbra before reaching the main cone. In some cases, the umbra is not entered at all. These purely penumbral lunar eclipses are quite unspectacular, but a slight dimming of the Full Moon may sometimes be detected with the naked eye. Penumbral lunar eclipses will occur on April 24, 2005 (mid-eclipse 09:55 UT) and March 14, 2006 (mid-eclipse 23:47 UT).

Over the next few years we will see the following umbral lunar eclipses (see overleaf):

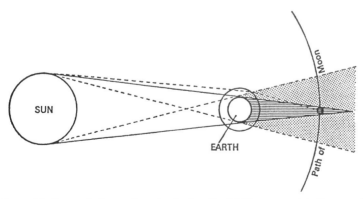

Figure 11. Theory of a lunar eclipse showing how the Full Moon moves into the cone of shadow cast by the Earth.

Date	Type	Mid-Eclipse (UT)		Duration of Totality	
		h	m	h	m
May 4, 2004	Total	20	31	1	16
Oct 28, 2004	Total	03	04	1	21
Oct 17, 2005	Partial, 7 per cent	12	04	–	
Sept 7, 2006	Partial, 19 per cent	18	52	–	
Mar 3, 2007	Total	23	21	1	16
Feb 21, 2008	Total	03	26	0	51
Aug 16, 2008	Partial, 81 per cent	21	10	–	

During a total lunar eclipse, the Moon does not (usually) vanish, because some sunlight is refracted (bent) on to the lunar surface by way of the Earth's atmosphere. As the blue light is mostly scattered out of this refracted light, the light reaching the Moon will have a predominantly reddish hue. The precise colour of the eclipsed Moon during totality, and whether the eclipse is bright or dark, depend upon the conditions in Earth's upper atmosphere. A scale for rating the colour and darkness of a lunar eclipse was worked out by the French astronomer André Louis Danjon:

0 Very dark. Moon almost invisible.
1 Dark grey or brownish colour; surface details barely identifiable.
2 Deep red or rusty red, with a dark patch in the centre of the umbra, but with brighter edges.
3 Brick red, sometimes with a brighter, yellowish border to the shadow.
4 Very bright coppery- or orange-red, with a bluish cast to the edge of the umbra.

Following the Leonid meteor showers of 2000 and 2001, some observers claimed to have seen bright flashes on the Moon which they attributed to meteoroid impacts. Obviously, a total eclipse of the Moon would be an ideal time to watch for phenomena of this sort. However, a typical meteoroid is a tiny object, and would not be expected to produce a flash visible over a distance approaching 385,000 kilometres (almost 240,000 miles). An object of larger size would be needed – of the size which leads to meteorite falls on Earth – and meteorite falls are not generally associated with the Leonids or any of the other major meteor showers.

The average length of the shadow cast by the Earth is about 1,384,000 kilometres (860,000 miles), so that at the distance of the Moon from the Earth the main shadow cone has a mean diameter of just over 9,200 kilometres (5,720 miles), which is considerably greater than the diameter of the Moon, 3,476 kilometres (2,160 miles). Consequently, at some eclipses, if the Moon passes centrally through the umbral shadow, totality may last for up to an hour and three-quarters. Totality on May 4, 2004 will last one hour and sixteen minutes, as the Moon passes to the south of the umbra's centre at maximum (Figure 12). The Moon enters the umbra at 18:49 UT, early on the evening of May 4, is total between 19:53 and 21:09 UT, and leaves the umbra at 22:13 UT. Observers in Britain, and indeed most of northern and western Europe, will see the Moon rise that evening already partially eclipsed, with mid-eclipse at 20:31 UT.

There are many lunar eclipse legends. In an old Scandinavian poem, the *Edda*, it is said that the monster Managarmer is trying to swallow the Moon, and is staining the air and the ground with blood. Apparently, the Orinoco Indians believed that the Moon turned red because it was angered by their laziness, so following an eclipse they redoubled their efforts in their fields!

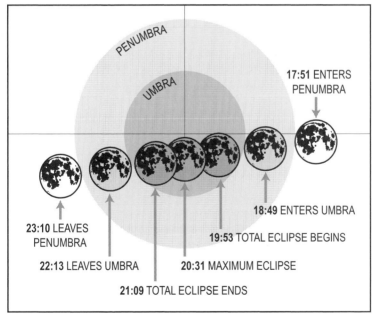

Figure 12. Path of the Moon through Earth's shadow on May 4, 2004.

June

New Moon: June 17   *Full Moon:* June 3

Solstice: June 21

MERCURY continues to be visible as a morning object, low above the east-north-eastern horizon for a short while before dawn, though only for the first week of the month, and also only for observers in equatorial and southern latitudes, who should refer to the diagram (Figure 10) given with the notes for May.

VENUS, magnitude −4.0, is now moving rapidly towards the Sun and will be seen in the evenings only for a short while after sunset, low in the western sky for the first couple of days of the month. For the second half of the month Venus may be seen as a morning object, magnitude −4.2, low above the eastern horizon before dawn. Venus passes rapidly through inferior conjunction on June 8, when it may be seen in a rare transit across the face of the Sun. The accompanying diagram (Figure 13) shows the apparent path of Venus as it crosses the Sun, as viewed from London and the Cape of Good Hope. Although the transit will be visible in its entirety (weather permitting) from London, the ingress will not be visible from the Cape. (For further information, see the notes about the transit of Venus on page 106 and 139 of this *Yearbook.*)

MARS is no longer a conspicuous object at magnitude +1.8, and observers in the latitudes of the British Isles will only be able to detect the planet low above the west-north-western horizon for a short while before the gathering twilight inhibits observation. These observers are unlikely to see it after the middle of the month. If they can see a faint point of light in that area it is more likely to be the star Pollux, which is 0.7 magnitudes brighter than Mars; the planet passes 5° south of the star around June 13–14. Observers in tropical and southern latitudes will continue to be able to view Mars throughout June.

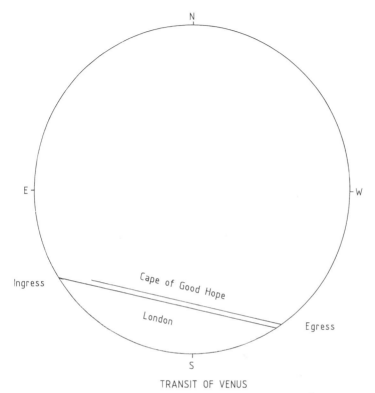

TRANSIT OF VENUS

Figure 13. The apparent track of Venus across the Sun during the transit of June 8, 2004, as viewed from London, UK, and from the Cape of Good Hope, South Africa.

JUPITER, magnitude −1.9, continues to be visible as a brilliant object in the western sky in the evenings.

SATURN, magnitude +0.1, is now coming to the end of its evening apparition, and is disappearing over the western horizon before dark. Observers in the latitudes of the British Isles will be lucky to see it even in the first few days of the month. In the Southern Hemisphere conditions are slightly better and the visibility period may last to the middle of the month.

PLUTO reaches opposition on June 11, in the constellation of Serpens, at a distance of 4,457 million kilometres (2,769 million miles) from the

Earth. It is only visible with a moderate-sized telescope since its magnitude is +14.

The Transit of Venus. Of course, the main event this month is the transit of Venus. There can be no living person who can remember the last, which was on December 6, 1882. This year's transit will be followed by another just eight years later, on June 6, 2012, and then there will be no more until December 11, 2117, an interval of 105 years.

There will be many people very anxious to see this month's transit. Clouds are always a potential menace on such occasions, and real enthusiasts may feel inclined to book a seat on an aircraft in case of an overcast sky, or else travel to a place where clouds are highly unlikely at this time of the year. The entire transit is visible throughout Europe except the extreme south-western Iberian Peninsula, Africa except the western part and extreme southern tip, Asia except for the extreme eastern part, most of the Indian Ocean, and the northern part of Greenland – so there are plenty of places from which to choose.

Since Venus's apparent diameter will be almost 58 arcseconds, compared with a disc diameter of about 1,890 arseconds for the Sun, Venus's silhouetted disc will be around ⅓₃ of the solar diameter. Consequently, during the transit Venus should be visible with the naked eye – but, as always, take the very greatest care. Do not stare at the Sun without using a solar filter that is safe for direct viewing; check it for scuffs, scratches or pinholes, and if you are in any doubt about the effectiveness of the filter then don't use it. On no account look directly through a telescope unless it is fitted with a solar filter that covers the telescope's full aperture and fits securely over the front end of the instrument. (Many small telescopes used to come supplied with a dark filter that fitted over the eyepiece. These are very dangerous, as they can crack under the magnified and focused heat of the Sun without warning. If you have one of these filters, please throw it away.) Much the best method for anyone who is unsure of the suitability of their equipment is to use a small telescope to project a magnified image of the Sun on to a white cardboard screen. Aim the instrument at the Sun using the shadow of the telescope tube on the card: never look through the main telescope or its smaller finder scope to do this.

The accompanying photograph of the 1882 transit (Figure 14) shows the size of Venus when seen in silhouette against the Sun. The phenomenon is no longer regarded as important, as it used to be in the

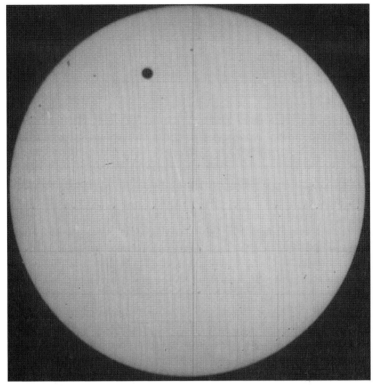

Figure 14. Venus is seen in silhouette against the brilliant face of the Sun in this image from a photographic plate showing the transit of Venus on December 6, 1882. (Image courtesy US Naval Observatory Library.)

days when it provided the best means of measuring the distance between the Earth and the Sun, but it will certainly be very interesting to watch. (For a fascinating historical account of past transits of Venus, read the article by Allan Chapman elsewhere in this *Yearbook*.)

The Status of Pluto. There has been a great deal of discussion about the status of Pluto, which comes to opposition this month. Pluto was discovered by a young astronomer, Clyde Tombaugh, using the 13-inch refractor at the Lowell Observatory in Flagstaff, Arizona, on photographic plates taken on January 23 and January 26, 1930, although the announcement was delayed until March 13, 1930 – 149 years after the discovery of Uranus and seventy-eight years after Percival Lowell's birth. The position of Pluto at discovery, near the star Delta Geminorum, was in excellent agreement with the prediction by Percival Lowell himself, who had worked out a position for his so-called 'Planet X' from its perturbations of Neptune and particularly Uranus; at that time, Pluto was believed to have a mass almost seven times that of the Earth. Lowell's predictions had certainly been accurate, but was it sheer luck? We now know that Pluto is far too small and lightweight to exert any significant perturbations upon giant planets such as Uranus and Neptune; current estimates give its mass as only about $\frac{1}{450}$ that of the Earth, or less than one-fifth the mass of our Moon.

Officially, Pluto is still classed as a planet, but the discovery of the Trans-Neptunian object Quaoar, whose diameter is more than half that of Pluto, alters the situation somewhat. Since 1992, astronomers have discovered several hundred Trans-Neptunian objects, with diameters greater than one hundred kilometres, in a ring or belt which extends outwards from the orbit of Neptune, at 30 AU from the Sun, to beyond 50 AU. This is generally known as the Kuiper–Edgeworth Belt.

Apart from Pluto (diameter 2,320 kilometres) and its moon Charon (diameter 1,270 kilometres), there are several other large Trans-Neptunian objects with diameters around the 1,000-kilometre mark; Quaoar (1,200 kilometres), Ixion 1,065 kilometres and Varuna (900 kilometres). Another object, 2002 AW197, is probably of comparable size. Accordingly, it is starting to look very much as though Pluto should be regarded merely as the senior member of the Kuiper–Edgeworth Belt. It may well be that many other similar-sized bodies or even larger bodies lie further out in the Solar System. But for the present, and probably until a larger body is found, Pluto retains its status as the ninth planet.

July

New Moon: July 17   *Full Moon:* July 2 and 31

EARTH is at aphelion (farthest from the Sun) on July 5 at a distance of 152 million kilometres (94 million miles).

MERCURY reaches greatest eastern elongation (27°) on July 27. The long duration of twilight, together with its southerly declination, renders it unobservable to those in the latitudes of the British Isles. Further south, in equatorial and southern latitudes it is visible as an evening object throughout the month. For observers in southern latitudes this will be the most favourable evening apparition of the year. Figure 15 shows, for observers in latitude 35°S, the changes in azimuth (true bearing from the north through east, south and west) and altitude of Mercury on successive evenings when the Sun is 6° below the horizon. This condition is known as the end of evening civil twilight and in this latitude and at this time of year occurs about thirty minutes after sunset. The changes in the brightness of the planet are indicated by the relative sizes of the circles marking Mercury's position at five-day intervals. It will be noticed that Mercury is at its brightest before it reaches greatest eastern elongation (27°) on July 27. Mercury overtakes Mars during the night of July 10–11, both planets moving eastwards. At closest approach, Mars will be seen only 10 arcminutes south of Mercury, which is about six times brighter than Mars. This phenomenon will only be visible to observers in equatorial and southern latitudes, and they are advised to use binoculars.

VENUS is a brilliant object, magnitude −4.5, dominating the eastern sky in the mornings before dawn. It attains its greatest brilliancy on July 15. Venus is pulling rapidly away from the Sun and by the end of the month is visible for about two hours before sunrise.

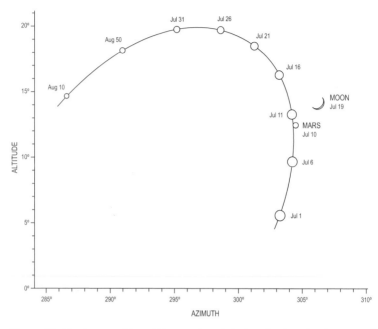

Figure 15. Evening apparition of Mercury, from latitude 35°S. (Angular diameters of Mercury, the crescent Moon and Mars not to scale.)

MARS, magnitude +1.8, is still visible to Southern Hemisphere observers, low in the western sky after darkness has fallen, but by the end of the month it will be a difficult object to detect.

JUPITER, although visible throughout the month as a conspicuous object in the western sky in the evenings, is nearing the end of its evening apparition and will not be visible for long after sunset by the end of the month for observers in the latitudes of the British Isles. The period available for observation lengthens as one moves further south. Jupiter's magnitude is −1.8.

SATURN passes through conjunction on July 8 and is therefore unsuitably placed for observation.

Intruder Into the Zodiac. Tradition says that there are twelve constellations of the Zodiac. In fact, there are thirteen, since Ophiuchus, the Serpent-Bearer, intrudes into the belt for some distance between Scorpius and Sagittarius. Astrologers, quite understandably, have an ill-concealed hate of Ophiuchus, and simply do not know what to do about him!

In fact, Ophiuchus, though large (Figure 16), is not a particularly prominent constellation. Its brightest star is Alpha (Rasalhague) magnitude 2.1; then come Eta (Sabik) 2.4, Zeta (Han) 2.6, Delta (Yed Prior) 2.7 and Beta (Chelab) 2.8. Of these, Zeta, with a luminosity 5,000 times that of the Sun, is much the most luminous; it is over 550 light years away.

In mythology Ophiuchus (formerly called Serpentarius) commemorates Æsculapius, son of the god Apollo, who became so skilled in medicine that he was able to restore the dead to life. To avoid depopulation of the Underworld, Jupiter reluctantly disposed of Æsculapius with a thunderbolt, but relented sufficiently to place him in the sky. He is shown struggling with Serpens (the Serpent), and seems

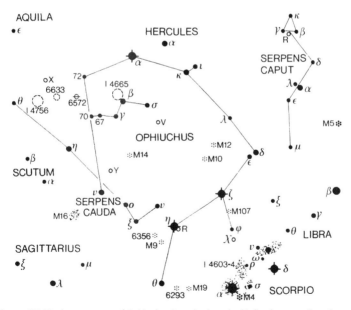

Figure 16. The large pattern of Ophiuchus intrudes into the Zodiac between Scorpius and Sagittarius. The snake, with which Ophiuchus struggled, is in two parts. Its head (Serpens Caput) lies to one side of Ophiuchus, and its body (Serpens Cauda) to the other.

to be having the better of matters, since the luckless reptile has been pulled in half. The planet Pluto lies in Serpens at the present time.

In 1777, the astronomer Poczobut attempted to form a new constellation, Taurus Poniatowski (Poniatowski's Bull), out of faint stars in the region of 70 Ophiuchi. The constellation was soon rejected. The star 70 Ophiuchi is a wide, easy binary, magnitudes 4.2 and 6.0, separation 3.4 arcseconds, period 88.1 years.

There are several globular clusters in Ophiuchus, of which seven are included in Messier's catalogue – M9, M10, M12, M14, M19, M62 and M107; they are all worth seeking out. In Serpens, the head (Caput) is more conspicuous than the body (Cauda); the red Mira variable R Serpentis, between Beta (magnitude 3.7) and Gamma (3.8) can reach the fifth magnitude at maximum. Its period is 358 days – little more than a week short of a year – so that there are periods when the star is too near the Sun when at maximum for it to be seen. In Cauda, Theta or Alya is a very wide double; magnitudes 4.5 and 4.5, separation 22.4 arcseconds. Any small telescope will show these 'identical twins'. Each component is twelve times as luminous as the Sun and is of spectral type A5. The distance from Earth is 102 light years.

The First Successful Ranger Probe. Between 1961 and 1965, the Americans launched their series of Ranger probes, their first attempt to obtain close-up images of the lunar surface. The Ranger spacecraft were designed to crash into the Moon and to send back valuable data and images until the moment of impact. It must be admitted that the early Ranger probes were not a success: the launch vehicle failed on Rangers 1 and 2; Ranger 3 missed the Moon by 37,000 kilometres and sent back no images; Ranger 4 landed on the night-side due to an instrument and guidance failure; and Ranger 5 missed the Moon by 630 kilometres and no data were received. Finally, in January 1964, Ranger 6 reached the Moon, coming down in the Mare Tranquillitatis, but the camera failed and no pictures were received. There was success at last forty years ago this month; on July 31, 1964, Ranger 7 crash-landed in the Mare Nubium (the Sea of Clouds), and 4,308 pictures were returned. Ranger 7 impacted in a region of lunar mare terrain modified by crater rays. Its first image was taken about seventeen minutes before impact.

August

New Moon: August 16   *Full Moon:* August 30

MERCURY is visible as an evening object, low above the west-south-western horizon, during the first week of the month, though only to observers in equatorial and southern latitudes. During this week its magnitude fades from +0.6 to +1.3. Mercury passes rapidly through inferior conjunction on August 21, but does not become visible as a morning object until September.

VENUS, magnitude −4.3, is a magnificent object in the eastern sky before sunrise, reaching its greatest western elongation (46°) on August 17. On the last day of the month, Saturn, which has been moving outwards from the Sun, passes 2° north of the very much brighter inner planet.

MARS is now too close to the Sun for observation and will not be seen again until November.

JUPITER, magnitude −1.7, continues to be visible as a bright evening object low in the western sky after sunset. Observers in tropical and southern latitudes will be able to see it throughout August, but observers further north will have the period of observation curtailed, and those in the latitudes of the British Isles are unlikely to see the planet after the first few days of the month.

SATURN, magnitude +0.2, is slowly emerging from the morning twilight, becoming visible low above the eastern horizon. It is moving slowly eastwards in the constellation of Gemini and its position among the stars is shown in Figure 1, given with the notes for January.

URANUS, magnitude +5.7, is barely visible to the naked eye, though it is readily located with only a small optical aid. The accompanying diagram (Figure 17) shows the path of Uranus against the stars during

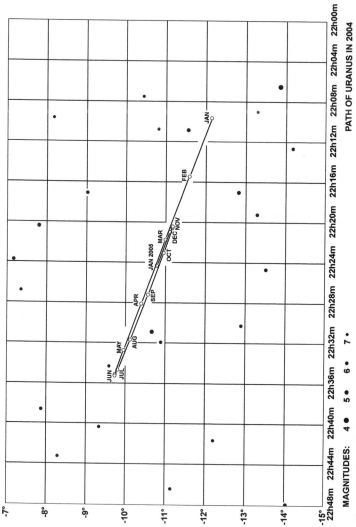

Figure 17. The path of Uranus against the stars of Aquarius during 2004. The fairly bright star at lower right is Iota Aquarii.

the year. The brightest star, magnitude +4.4, is Iota Aquarii, shown at a position of RA 22h 06.7m, dec. −13.85°. At opposition on August 27, Uranus is 2,849 million kilometres (1,770 million miles) from the Earth.

NEPTUNE is in the constellation of Capricornus and comes to opposition on August 6. It is not visible to the naked eye since its magnitude is +7.8. The accompanying diagram (Figure 18) shows the path of Neptune against the stars during the year. At opposition on August 6, Neptune is 4,346 million kilometres (2,700 million miles) from the Earth.

The Outer Giants. The four large planets – Jupiter, Saturn, Uranus and Neptune – are often classed together, but this is misleading. The Jupiter/Saturn pair is very different from the Uranus/Neptune pair, and it may be best to regard Jupiter and Saturn as gas giants, and Uranus and Neptune as ice giants. Moreover, there are marked differences between Uranus and Neptune. Neptune is very slightly the smaller of the two, but it is decidedly more massive, and it has a marked internal heat source, which Uranus does not. This means that the temperatures of the outer atmospheres of the two outer giants are almost exactly the same (around −215°C.), though Neptune is much further from the Sun.

Uranus is a relatively bland world; Neptune is much more dynamic. Both were passed by the Voyager 2 probe (Uranus in January 1986, Neptune in August 1989), and excellent images were sent back. Neptune showed a Great Dark Spot, but more recent images obtained with the Hubble Space Telescope indicate that this spot has disappeared.

Both Uranus and Neptune have magnetic fields, but the magnetic axes are nowhere near their axes of rotation, and do not even pass through the centres of the planets. When this was established for Uranus, it was suggested that the planet might be undergoing a magnetic reversal, but this seems to have been disproved by the fact the same situation applies to Neptune; two magnetic reversals at the same time would be too much of a coincidence. Both ice giants have obscure ring systems.

Ordinary telescopes show no surface details on Uranus or Neptune, but their dim, pale discs are easy to identify. Uranus is greenish, Neptune bluish. Their satellite systems are extensive: as of April 2003

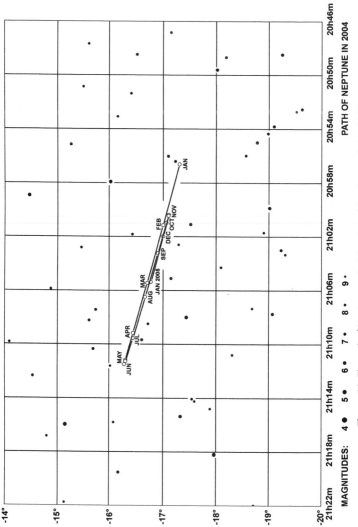

Figure 18. The path of Neptune against the stars of Capricornus during 2004.

Uranus is known to have twenty-one moons, and Neptune eleven, though most of these are small and some are probably asteroidal. Triton, the senior satellite of Neptune, was imaged from Voyager 2; it has an extremely tenuous atmosphere and is made up of a mixture of rock and ice. There is a surface coating of ice, presumably water ice overlaid by nitrogen and methane ices; normal craters are rare, and there is little surface relief. Triton's south polar cap is coated with pink nitrogen snow and ice, and there are active nitrogen geysers.

Perseus. The Perseid meteors, so named because the radiant (the region in the sky from where the meteors appear to emanate) is in Perseus, are always reliable. The shower peaks on around August 12 and lasts for some weeks – between late July and the third week of August – and this year, with New Moon on August 16, moonlight will not be a problem.

Perseus itself is a distinctive constellation (Figure 19), even though it contains no first-magnitude star; its leader is Alpha (Mirphak), magnitude 1.8. The most celebrated star in Perseus is, of course, the eclipsing binary Algol, the 'Demon Star', which is normally of magnitude 2.1, but

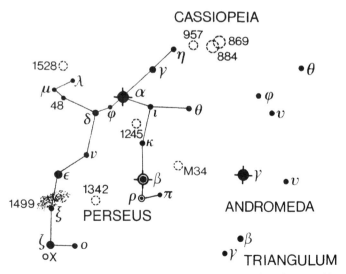

Figure 19. The principal stars of Perseus, showing the locations of the three variable stars Algol (Beta Persei), Rho and X Persei, the open cluster M34, and the rich star clusters of the Sword Handle, NGC 869 and 884.

fades to 3.4 at minima, which occur every 2.867 days, or 68.8 hours. The main component is a hot, blue B-type star, about three and a half times the mass of the Sun; the secondary is a slightly larger but less massive orange star of type K, with a mass about the same as our Sun. Algol fades to magnitude 3.4 when the dimmer K-type star passes in front of the B-type star as seen from Earth; the eclipses are not total and last only a few hours (Figure 20). It also fades (but only by 0.1 magnitude) when the K-type passes behind the B-type star. Algol is, of course, brightest when you have an unobstructed view of both stars.

Algol's fluctuations are easy to follow with the naked eye. Near it is Rho Persei, an M-type semi-regular variable, with a range from magnitude 3.3–4.0, and a very rough period of around fifty days. Forming a triangle with Algol and Rho is the open star cluster M34, which is just visible with the naked eye under good conditions.

The Sword Handle in Perseus is magnificent; two rich star clusters in the same low-power field. Their NGC numbers are 869 and 884. They were not listed by Messier, presumably because there was no fear of confusing them with a comet.

X Persei, close to Zeta, is an X-ray source. It is classed as a Gamma Cassiopeiae star, and is variable between magnitudes 6 and 7.

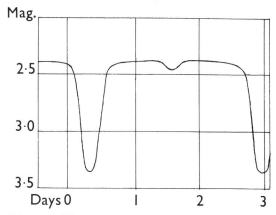

Figure 20. Light curve of the eclipsing binary star Algol, the 'Demon Star', showing the pattern of primary and secondary eclipses; the primary eclipses recur at intervals of 68.8 hours.

September

New Moon: **September 14**   *Full Moon:* **September 28**

Equinox: September 22

MERCURY is visible as a morning object for a period of nearly three weeks, after the first few days of the month, for observers in northern and equatorial latitudes. For observers in northern temperate latitudes this will be the most favourable morning apparition of the year. Figure 21 shows, for observers in latitude 52°N, the changes in azimuth (true bearing from the north through east, south and west) and altitude of Mercury on successive mornings when the Sun is 6° below the horizon. This condition is known as the beginning of morning civil twilight and in this latitude and at this time of year occurs about thirty-five minutes before sunrise. The changes in the brightness of the planet are indicated by the relative sizes of the circles marking Mercury's position at five-day intervals. It will be noticed that Mercury is at its brightest after it reaches greatest western elongation (18°) on September 9.

VENUS continues to be visible as a brilliant object in the eastern morning sky before dawn. Its magnitude is −4.2.

MARS passes through conjunction on September 15 and therefore remains unsuitably placed for observation.

JUPITER, magnitude −1.7, has come to the end of its evening apparition, though given near perfect conditions it may just be possible for observers in southern latitudes to glimpse it low in the western sky just after sunset for the first few days of the month. Jupiter passes through conjunction on September 21.

SATURN continues to be visible as a morning object in the eastern sky, magnitude +0.2. It becomes visible rising above the east-north-east horizon before midnight, by the end of the month, for observers in

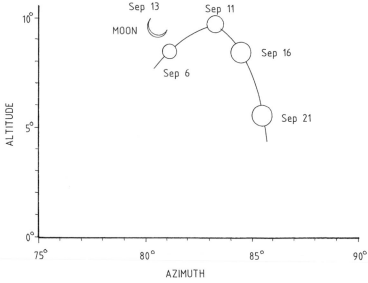

Figure 21. Morning apparition of Mercury, from latitude 52°N. (Angular diameters of Mercury and the crescent Moon not to scale.)

northern temperate latitudes, but several hours later for southern hemisphere observers.

Deneb and the Northern Cross. Deneb, one of the members of the Summer Triangle, is almost overhead during September evenings from the latitudes of Britain. (The term 'Summer Triangle', comprising Vega in Lyra, Deneb in Cygnus and Altair in Aquila, is entirely unofficial, but is widely used, even though it is not suited to the Southern Hemisphere.)

Deneb, or Alpha Cygni, is the brightest star in Cygnus (Figure 22), which is a splendid constellation; officially it represents a Swan but is often called the Northern Cross, for obvious reasons – it is far more cruciform than the Southern Cross. One member of the pattern, Albireo or Beta Cygni, is fainter than the rest but it is probably the loveliest coloured double star in the sky, with a golden-yellow primary (magnitude 3.1) and a vivid blue companion (5.1). The pair are 34.3 arcseconds apart, and any small telescope will separate them.

How bright a star appears to us in the sky (its apparent magnitude) depends upon a number of factors: how luminous the star is intrinsically, how far away the star is from us, and how much dust there is between us and the star. One way to put all stars on a common scale is to remove the effect of their varying distances from us by comparing their absolute magnitudes rather than their apparent magnitudes. The absolute magnitude of a star is the apparent magnitude that it would have if seen from a standard distance of 32.6 light years (10 parsecs).

The absolute magnitude of Deneb is −8.7, so that if seen from a distance of just 32.6 light years it would cast shadows. The absolute magnitude of our Sun is +4.8, so it is clear that Deneb is intrinsically an exceptionally luminous star – more than 250,000 times as luminous as the Sun.

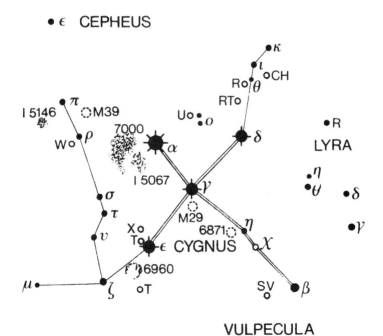

Figure 22. Deneb is the brightest star in the pattern of Cygnus, the Swan, which is also known as the Northern Cross. The remarkable variable star P Cygni lies just to the south-west of Gamma Cygni, which is at the intersection of the two arms of the 'Cross'.

A rough link between absolute magnitude and luminosity is given below:

Absolute Magnitude	Luminosity (Sun = 1)
−9	331,000
−8.5	209,000
−8	132,000
−7.5	83,200
−7	52,500
−6.5	33,100
−6	20,900
−5.5	13,200
−5	8,300
−4.5	5,250
−4	3,300
−3.5	2,100
−3	1,300
−2	525
−1	210
0	83
+1	33
+2	15
+3	5.3
+4	2.1
+5	0.9
+6	0.3

A Remarkable Variable Star. Among the stars of Cygnus, Deneb is not the only 'searchlight'. The unusual variable star P Cygni, which lies just to the south-west of Gamma Cygni or Sadr, is also exceptionally luminous. P Cygni has an estimated absolute magnitude of around −9 (although there is some dimming by interstellar dust), making it at least 330,000 times as luminous as our Sun, and one of the most luminous stars in our Galaxy. Its distance is given as around 6,000 light years.

When P Cygni rose to magnitude 3, in 1600, it was regarded as a nova, since it appeared where no star had been seen before. This outburst went on for a few years, before the star faded to below naked-eye visibility. It rose again to the third magnitude in 1655, where it remained for several years, until it declined once more to below magnitude 6. P Cygni brightened again in 1665 and, after fluctuating erratically in brightness, it steadied at about magnitude 5 in the early eighteenth century. For many years now it has hovered between magnitudes 4.8 and 5.2, just visible to the naked eye under clear, dark conditions, and an easy binocular object. P Cygni's brightness bursts are now explained, as in the case of certain other luminous blue variables, by the assumption that the star is ejecting vast amounts of matter. It is surrounded by a faint nebula and by shells of gas that are a consequence of past eruptions. The star is now classed as a variable of the S Doradûs type.

October

Summer Time in the United Kingdom ends on October 31.

MERCURY passes through superior conjunction on October 5, and it is not until the last week of the month that it becomes visible low in the western sky in the evenings to observers in tropical and southern latitudes, magnitude -0.5. It will not be visible to observers as far north as the British Isles.

VENUS is still a brilliant object in the eastern morning sky before sunrise. Its magnitude is -4.1.

MARS continues to be too close to the Sun for observation.

JUPITER, magnitude -1.7, is emerging from the morning twilight and will be seen low above the eastern horizon before dawn. Observers in the latitudes of the British Isles may hope to see the planet after the first week of the month, but those in southern latitudes will have to wait until almost the end of October before conditions are suitable for observation. Jupiter is in the constellation of Virgo.

SATURN, magnitude $+0.1$, is now visible in the eastern sky in the late evenings for observers in the latitudes of the British Isles, though later still as we move into southern latitudes. During the month Saturn moves from Gemini into Cancer. The observer armed with a small telescope will notice that the rings of Saturn present a beautiful spectacle. Early in 2003, the rings were at their maximum opening, when the rings extended beyond the polar diameter of Saturn. They are now closing slightly, and this month the polar diameter of the minor axis of the rings is 15 arcseconds, slightly less than the polar diameter of the planet itself.

The Next Galactic Supernova? The last supernova to have been definitely observed in our Galaxy flared up 400 years ago this month, close to the star Lambda Ophiuchi in the constellation of the Serpent-Bearer. It was discovered on October 9, 1604, and the great German astronomer Johannes Kepler first saw it on October 17, but he followed its brightness variations so carefully that it is always remembered as Kepler's Star. His book on the subject was entitled *De Stella Nova in Pede Serpentarii* (On the new star in Ophiuchus's foot). It rose to a peak magnitude of −2.5 in late October, and still rivalled Jupiter in brilliance when it was lost in the evening twilight in November. On its reappearance in January 1605, Kepler noted that it was brighter than Antares (magnitude 1.0), and it remained visible with the naked eye until March 1606.

From the light curve, Kepler's Star is thought to have been a Type 1a supernova, probably caused by the explosion of a white dwarf in a binary system as a consequence of matter falling on to it from its companion star. Kepler's Star left a nineteenth-magnitude remnant, a faint fan-shaped nebulosity about 40 arcseconds in extent, consisting of some faint, wispy filaments and bright condensations. It is identified with the radio source 3C 358, and lies at distance of around 20,000 light-years.

There may have been a supernova in Cassiopeia around 1667, leaving a radio source Cassiopeia A (Figure 23), but it was not seen because it was so heavily obscured by interstellar dust near the main plane of the Galaxy. We have, of course, seen the supernova of 1987 in the nearby Large Magellanic Cloud, but can we tell when another of these colossal outbursts will occur in our own Galaxy? We cannot, but there are two very promising candidates: Eta Carinae and Rho Cassiopeiae.

Eta Carinae, in the keel of the old ship Argo Navis, is too far south to be seen from Europe. For many years now, it has hovered on the brink of naked-eye visibility; most of its radiation is in the infrared part of the spectrum. But at one time, in April 1843, Eta Carinae attained magnitude −0.8, outshining every star in the sky apart from Sirius; though the star released as much visible light as a supernova explosion, it survived the outburst. The explosion produced two lobes and a large, thin equatorial disc (Figure 24), all moving outwards at about 2.4 million kilometres per hour (1.5 million miles per hour). Clearly it is a very unstable star, of exceptional mass (estimated at 100 solar masses), and

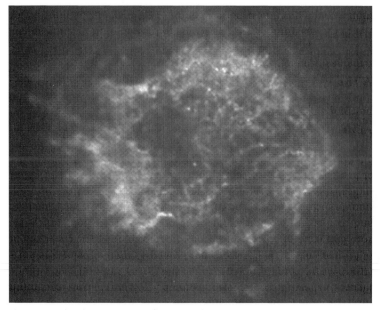

Figure 23. Chandra X-ray image of the 'young' supernova remnant Cassiopeia A. (Image courtesy NASA/CXC/SAO/Rutgers University/John Hughes.)

associated with nebulosity; the mass is in fact so great that some astronomers believe that we are dealing with a binary system rather than a single exceptional star. Eta Carinae's mass makes it an excellent candidate for a supernova. It is estimated to lie at a distance of at least 8,000 light years.

The other prime candidate, Rho Cassiopeiae, is circumpolar in Britain. It lies near the famous 'W' of Cassiopeia (Figure 25), between the stars Tau Cassiopeiae (magnitude 4.9) and Sigma (also 4.9), which act as useful comparison stars for the binocular observer. Usually Rho is about magnitude 4.8 but sometimes, as in 1945–46, 1985–86 and 2000–2001, it fades briefly to magnitude 6. Rho is a hypergiant – far more powerful than a normal supergiant. Radiating more than half a million times more light than the Sun, Rho has a diameter between 400 and 500 times that of the Sun. Its distance is at least 8,000 light years and may be as much as 10,000 light years. Hypergiants such as Rho Cassiopeiae are scarce; less than one in a million stars in our Galaxy is a hypergiant. Like Eta Carinae, Rho is highly unstable; it is ejecting about

Figure 24. A huge, billowing pair of gas and dust clouds is captured in this stunning Hubble Space Telescope image of the super-massive star Eta Carinae. Even though Eta Carinae is more than 8,000 light years away, features sixteen billion kilometres (ten billion miles) across – about the diameter of our Solar System – may be distinguished. (Image courtesy Jon Morse, University of Colorado, and NASA/STScI.)

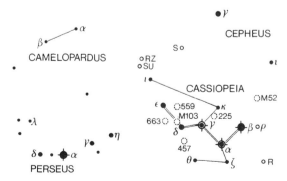

Figure 25. The familiar 'W' of Cassiopeia showing the positions of the interesting variable stars Rho and Gamma Cassiopeiae. Alpha Cassiopeiae is also suspected of variability.

$^1/_{20}$ of a solar mass of its gas and dust annually. There is no doubt that it will explode as a supernova; the only question is when? It may not be for centuries, but it could be quite soon.

During the recent 2000–2001 eruption, Rho first brightened briefly as gases fell in towards the star and were compressed and heated, then some of that material was blasted outwards in a powerful shock wave, causing the star to fade from magnitude 4 to 6. At the same time, its spectral type changed from type F to type M, indicating a drop in surface temperature. Since then, the star's atmosphere has been pulsating in a strange manner, and an even larger eruption may be imminent.

In the W of Cassiopeia, Gamma is an interesting variable. It is usually just below the second magnitude but can sometimes rise to 1.6, or fade to below 3. Alpha Cassiopeiae (Shedir) is of type K and is definitely orange-red. It is suspected of slight variability. Beta (magnitude 2.3) is a suitable comparison for both Alpha and Gamma; also use Delta (2.7).

Another Total Eclipse of the Moon. The second total lunar eclipse of 2004 takes place on the night of October 27–28. Western Europe (including the British Isles), West Africa, South America and most of North America will see this total eclipse in its entirety. In the far western parts of North America, the Moon will rise already eclipsed, and in Eastern Europe, Central and East Africa, the Middle East and most of Asia, the Moon will set before the eclipse ends. On this occasion, the Moon passes north of the centre of the umbra; the Moon enters the umbra at 01.15 UT, early on the morning of October 28, is total between 02:24 and 03:45 UT, and leaves the umbra at 04:54 UT.

November

New Moon: November 12   *Full Moon:* November 26

MERCURY is an evening object throughout the month, though not for observers in latitudes as far north as the British Isles. Further south, and particularly in tropical and southern latitudes, Mercury may be seen in the early evenings, low above the west-south-western horizon, about half an hour after sunset. During the month its magnitude fades from −0.4 to +0.5. The planet will be at its brightest before it reaches greatest eastern elongation (22°) on November 21.

VENUS, magnitude −4.0, continues to be visible as a brilliant object in the eastern sky before dawn. Venus and Jupiter are in the same area of sky in the first part of the month, and on November 5 Venus passes about 0.5° north of Jupiter.

MARS, magnitude +1.7, becomes a morning object early in the month, low above the south-eastern horizon, by about 06 hours for observers in the latitudes of the British Isles. Gradually the visibility area extends further south until by the end of the month it extends to observers in southern temperate latitudes. Figure 26 shows the path of Mars among the stars for November–December.

JUPITER continues to be visible in the eastern sky as a brilliant morning object, magnitude −1.7. It is moving outwards from the Sun and steadily becoming visible for longer and longer each morning. By the end of November it is visible for several hours before sunrise.

SATURN, magnitude 0.0, is still technically visible as a morning object, though appearing in the eastern skies in the latter part of the evening. Saturn is moving direct in the constellation of Cancer until it reaches its first stationary point on November 8, when it commences its retrograde motion.

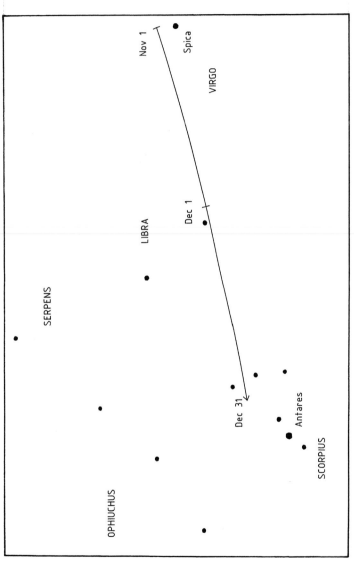

Figure 26. The path of Mars during November and December 2004.

The Craters of Mars. Forty years ago – on November 28, 1964 – the first successful Mars probe was launched. Mariner 4 flew past Mars on July 14, 1965 at a range of 9,846 kilometres (6,118 miles) and sent back the first pictures which showed unmistakable craters on Mars.

It was not the first attempt. Russia's Mars 1, launched on November 1, 1962, had undoubtedly encountered Mars, but all contact with it had been lost much earlier, and no scientific data were obtained. The American probe Mariner 3 (launched November 5, 1964) was a total failure, and all contact was lost soon after launch. However, Mariner 4 more than made up for these earlier setbacks, becoming the first spacecraft to get a close look at Mars. In addition to carrying various field and particle sensors and detectors, the spacecraft confirmed the thinness of the Martian atmosphere, and took twenty-two pictures covering about 1 per cent of the planet with its television camera. Initially stored on a four-track tape recorder, these pictures took four days to transmit to Earth. Mariner 4 remained in contact until December 21, 1967.

Up to that time, it had been thought that Mars was a world without major mountains or valleys, although many years earlier it had been claimed that craters had been seen from Earth: by E.E. Barnard, using the Lick Observatory 36-inch refractor (1892), and J. Mellish, using the Yerkes 40-inch refractor (1917). Unfortunately, their observations were never published. Craters had been predicted in 1944 by D.L. Cyr; his reasoning was completely wrong, but Mariner 4 showed that craters did exist, and that some were very large indeed.

By sheer bad luck, Mariner 4 and the next two successful Mars probes, Mariners 6 and 7 (1969), surveyed the least interesting parts of Mars. It was only when Mariner 9 was put into a closed path around the planet, in November 1971, that we had our first views of the valleys and the volcanic peaks. Moreover, excellent images were obtained of the polar caps – then widely believed to be composed mainly of carbon dioxide ice. Not until 2003 was it finally shown that the caps consist almost entirely of water ice with only a thin coating of carbon dioxide ice.

The Mariner 4 pictures (Figure 27) seem very poor by today's standards, but at the time they represented a tremendous technical triumph, and paved the way for all that followed. We have learned a great deal about Mars in the past forty years.

Figure 27. Mariner 4 image showing at least twenty craters of various sizes in the western Memnonia Fossae region of Mars. The image was taken from a range of 13,000 kilometres and measures 253 by 225 kilometres. (Image courtesy NASA/NSSDC.)

New Names for Planetary Satellites. Many new moons have been found recently around the giant planets of our Solar System, and some of these have been officially named by the Working Group on Planetary System Nomenclature of the International Astronomical Union.

Satellites of Uranus have been generally named after Shakespearian characters – a departure from the usual mythology, and something that does not meet with universal approval. The new outer irregular satellites found in 1997 were named Caliban (XVI, S/1997 U1) and Sycorax (XVII, S/1997 U2). We now have Prospero (XVIII, S/1999 U3), Setebos (XIX, S/1999 U1) and Stephano (XX, S/1999 U2); all these are small. Another two of Uranus's current (April 2003) total of twenty-one moons have yet to be named.

The names of eleven of Jupiter's new satellites discovered in 1999 and 2000 follow the usual mythological pattern:

XVII	S/1999 J1	Callirrhoe
XVIII	S/2000 J1 (= S/1975 J1)	Themisto
XIX	S/2000 J8	Megaclite
XX	S/2000 J9	Taygete
XXI	S/2000 J10	Chaldene
XXII	S/2000 J5	Harpalyke
XXIII	S/2000 J2	Kalyke
XXIV	S/2000 J3	Iocaste
XXV	S/2000 J4	Erinome
XXVI	S/2000 J6	Isonoe
XXVII	S/2000 J7	Praxidike

Note that S/2000 J1, Themisto, was finally recovered after a gap of twenty-five years. Callirrhoe and Themisto are the largest of the twelve moons discovered in 1999 and 2000, with diameters of about eight kilometres; all the others are only three to seven kilometres across. S/2000 J11, one of the eleven new moons discovered in 2000, has yet to be named.

With twenty-eight moons known at the end of 2000, the pace at which new Jovian satellites have been discovered has quickened in recent years; eleven new moons were discovered in 2001, one in 2002, and twenty in the first four months alone of 2003, bringing Jupiter's total number of known satellites (as of April 2003) to sixty! Given the sheer number of discoveries, it may be some time before names for all the new satellites are announced!

Saturn in Cancer. Cancer, the Crab, is one of the faintest of the Zodiacal constellations; Saturn now lies in it, and is much the brightest object in the area. Cancer lies inside the large triangle formed by the stars Pollux in Gemini, Regulus in Leo and Procyon in Canis Minor (Figure 28). Cancer represents the sea-crab which Juno, queen of Olympus, sent to the rescue of the multi-headed Hydra which was doing battle with Hercules. Not surprisingly, Hercules trod on the crab, but as a reward for its efforts Juno placed it in the sky. Only two stars in Cancer, Beta or Altarf (magnitude 3.5) and Delta or Asellus Australis (3.9), are above the fourth magnitude, but the pattern is redeemed by the presence of two bright open clusters: M44 or Praesepe, close to Gamma and Delta, and M67, which lies roughly one-quarter of the distance from Acubens or Alpha Cancri (4.2) towards Altarf. Praesepe is

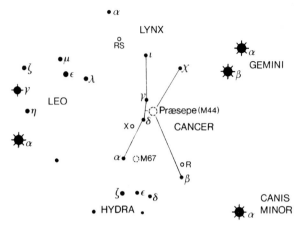

Figure 28. The rather obscure pattern of Cancer, the Crab, lies within the triangle formed by the first-magnitude stars Pollux (Beta Geminorum), Regulus (Alpha Leonis) and Procyon (Alpha Canis Minoris).

an easy naked-eye object, and one of the loveliest open clusters in the sky; it is difficult to understand why the Chinese referred to it as 'the exhalation of piled-up corpses'! M67, one of the oldest known galactic clusters, is on the fringe of naked-eye visibility. X Cancri, near Delta, is a very red semi-regular variable of spectral type C, range magnification 5.6–7.5, so that it is always within the range of binoculars. There is a rough period of about 195 days.

December

Solstice: December 21

MERCURY is a difficult evening object. Even for observers in tropical and southern latitudes it will only be visible for the first few days of the month. It may then be glimpsed low above the western horizon around the beginning of evening civil twilight, magnitude +1. Mercury passes rapidly through inferior conjunction on December 10. Shortly after the middle of the month, the planet becomes visible as a morning object to observers in equatorial and northern latitudes, and for the last ten days of the month for those in more southerly latitudes. It may then be seen low above the south-eastern horizon around the beginning of morning civil twilight. It reaches greatest western elongation (22°) on December 29. During this period of visibility its magnitude brightens from +1.1 to −0.3.

VENUS continues to be visible as a brilliant object, magnitude −4.0, in the eastern sky in the mornings before sunrise, though the period available for observation is shortening noticeably as it moves towards the Sun. On December 5, Venus passes 1° north of Mars, thus helping to identify this much fainter planet. Later in the month, on December 28–29, Venus passes 1° south of Mercury.

MARS, magnitude +1.6, is beginning to be more easily visible as a morning object, visible for some time in the south-eastern sky before the increasing brightness of the pre-dawn sky inhibits observation. By the end of the year Mars has approached to within 10° of Antares, the star being about 0.6 magnitudes brighter than the planet.

JUPITER, magnitude −1.9, is a brilliant morning object, in the south-eastern sky.

SATURN does not reach opposition until next month, but is becoming visible for most of the hours of darkness. Its magnitude is now −0.2. Saturn is moving slowly retrograde and returns from Cancer into Gemini during the month.

A Sad Beginning to 2003. The first couple of months of 2003 saw two notable disasters, neither of which could have been foreseen. On January 18, a tremendous bush fire swept across the Canberra region of Australia, virtually destroying the Mount Stromlo Observatory, and wiping out a third of Australia's world-leading astronomy programme. The flames destroyed five telescopes, the workshop, the homes of eight members of staff and the main dome. The lost telescopes included the 1.9-metre (74-inch) Grubb Parsons reflector, the 1.3-metre (50-inch) Great Melbourne Reflector built in 1868 and upgraded a decade ago, the 26-inch Yale–Columbia refractor, and the historic 9-inch Oddie refractor. The administration building housing the main library was also burned down, and another casualty was Mount Stromlo's workshop, which contained a 5-million-dollar imaging spectrograph being built for the Gemini North Telescope on Mauna Kea. In financial terms the loss amounts to more than twenty million dollars. Mercifully nobody was killed or injured, but the observatory staff had just twenty minutes to evacuate. The task of rebuilding will be a very long one.

Only a fortnight later, there was the tragedy of the Space Shuttle *Columbia*, which broke up on re-entering the Earth's atmosphere on February 1 with the loss of all seven crew members; fragments of the orbiter were strewn from Central Texas to Louisiana. Harold Gehman, Head of the Columbia Accident Investigation Board (CAIB), said his panel's report would examine every possible cause and flaw that might have led to the disintegration of *Columbia* during re-entry. The leading theory for the cause of the disaster was that a piece of insulating foam broke off the shuttle's external tank on lift-off, striking and damaging the critical leading edge of the orbiter's left wing. There appears to be some evidence that it *was* a failure of the left wing which led to the break-up of *Columbia*, but clearly we must await the panel's findings. The CAIB's report into the accident is not expected before the late summer of 2003 but, whether or not it is able to pinpoint the exact cause of failure, it is likely to take a very broad approach to the contributing factors in the disaster. Although the reason for the accident is not yet known – quite apart from the human tragedy – the human space programme is bound to be badly delayed.

William McCrea. Professor Sir William Hunter McCrea FRS died on April 25, 1999 at the age of 94. He was born in Dublin on December 13, 1904 – one hundred years ago this month – but was educated in England, first at Chesterfield Grammer School, and then at Trinity College, Cambridge, where he read Mathematics. At the age of twenty-five he was appointed Lecturer in Mathematics at Edinburgh University, but he quickly moved to a Readership at Imperial College, London. In 1936, he returned to Ireland as Chair of Mathematics at Queen's University, Belfast, and after war service he joined Royal Holloway College as Professor and Head of Department, where he remained for twenty-two years until 1966. His final move was to the University of Sussex, where he set up the Astronomy Centre.

'Bill' McCrea established an international reputation as both an original researcher and an excellent teacher, and made many outstanding contributions in mathematics, physics, astronomy, and the applications of relativity to cosmology. In astrophysics, his work covered many topics from cosmic rays, novae and stellar evolution to the origin of the Solar System. He also made notable contributions on the nature of comets and the astronomical conditions for ice ages on Earth. He was President of the Royal Astronomical Society between 1961 and 1963, and was knighted in 1985. Universally liked, as well as admired, he remained active until the last few months of his long life.

The Return of Orion. In the northern hemisphere, winter is back – and in the evening sky so is Orion, the Hunter. This is not only a magnificent constellation in itself, but its stars also act as an invaluable guide to all the neighbouring constellations (Figure 29). Extending a line through the three stars of Orion's Belt, towards the south-east, brings you to Sirius, the leading star of Canis Major; in the opposite direction, a line through the Belt stars directs you to Aldebaran in Taurus. A line from Bellatrix (Gamma Orionis), passing through Betelgeux (also called Betelgeuse) and curved slightly southwards, takes you to Procyon in Canis Minor. Another slightly curved path from Delta Orionis (the westernmost star of the Belt) through Betelgeux and onwards will find Pollux, the brighter of the Twins of Gemini. Finally, another line from Delta Orionis, extended through Bellatrix to the

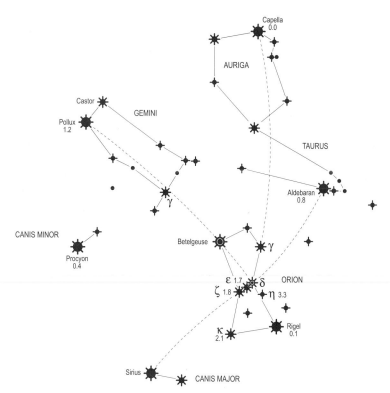

Figure 29. By extending imaginary lines through the principal stars of Orion, the pattern may be used as a signpost to help a beginner locate the bright stars of neighbouring constellations, and so learn their way around the night sky.

north, will eventually guide you to Capella, the principal star of Auriga. Use Orion as a signpost, as shown here, and it will not be long before you have built up a good working knowledge of the winter night sky.

The Transit of Venus in 2004

A rare transit of Venus across the face of the Sun occurs on June 8. The ingress only is visible from Australasia and the extreme part of eastern Asia. The complete transit is visible from the rest of Asia, Europe, Iceland and Greenland, the extreme northern part of North America and most of Africa apart from the west and extreme south-west. Only the egress will be visible from the remainder of the Americas, except the western part of North America and southern South America.

Geocentric times and position angles are as follows:

	h	m	°
Ingress, exterior contact	5	13.5	116
Ingress, interior contact	5	32.8	119
Egress, interior contact	11	06.5	213
Egress, exterior contact	11	25.9	216

A diagram is given with the monthly notes for June.

The ingress times for Australia and eastern Asia are several minutes earlier than those given above.

For London the ingress times will be about six or seven minutes later, and egress times about two minutes earlier than those given above.

The egress times for North America are within a few minutes of those given above.

Eclipses in 2004

During 2004 there will be four eclipses, two of the Sun and two of the Moon.

1. *A partial eclipse of the Sun on April 19* is visible from a small part of Antarctica, the South Atlantic Ocean, southern Africa, and Madagascar. The eclipse begins at 11^h 30^m and ends at 15^h 39^m.

2. *A total eclipse of the Moon on May 4* is visible from Australasia, the western Pacific Ocean, Antarctica, the Indian Ocean, Asia (except the extreme north-east), Europe, Africa, the Atlantic Ocean, Newfoundland and South America. The partial eclipse begins at 18^h 48^m and ends at 22^h 12^m. Totality lasts from 19^h 52^m to 21^h 08^m.

3. *A partial eclipse of the Sun on October 14* is visible from eastern and north-eastern Asia, western Alaska, and the North Pacific Ocean including Hawaii. The eclipse begins at 00^h 55^m and ends at 05^h 04^m.

4. *A total eclipse of the Moon on October 28* is visible from the western Indian Ocean, western Asia, Europe, Africa, the Atlantic Ocean, Greenland, the Arctic Ocean, extreme north-east Asia, the Americas and the Pacific Ocean. The eclipse begins at 01^h 14^m and ends at 04^h 54^m. Totality lasts from 02^h 23^m to 03^h 44^m.

Occultations in 2004

In the course of its journey round the sky each month, the Moon passes in front of all the stars in its path, and the timing of these occultations is useful in fixing the position and motion of the Moon. The Moon's orbit is tilted at more than 5° to the ecliptic, but it is not fixed in space. It twists steadily westwards at a rate of about 20° a year, a complete revolution taking 18.6 years, during which time all the stars that lie within about 6½° of the ecliptic will be occulted. The occultations of any one star continue month after month until the Moon's path has twisted away from the star, but only a few of these occultations will be visible from any one place in hours of darkness.

There are ten occultations of bright planets in 2004; two of Mercury, two of Venus, four of Mars, and two of Jupiter.

Only four first-magnitude stars are near enough to the ecliptic to be occulted by the Moon: these are Aldebaran, Regulus, Spica and Antares. However, none of these stars undergoes occultation in 2004.

Predictions of these occultations are made on a worldwide basis for all stars down to magnitude 7.5, and sometimes even fainter. The British Astronomical Association has produced a complete lunar occultation prediction package for personal-computer users.

Occultations of stars by planets (including minor planets) and satellites have aroused considerable attention.

The exact timing of such events gives valuable information about positions, sizes, orbits, atmospheres and sometimes of the presence of satellites. The discovery of the rings of Uranus in 1977 was the unexpected result of the observations made of a predicted occultation of a faint star by Uranus. The duration of an occultation by a satellite or minor planet is quite small (usually of the order of a minute or less). If observations are made from a number of stations it is possible to deduce the size of the planet.

The observations need to be made either photoelectrically or visually. The high accuracy of the method can readily be appreciated when one realizes that even a stopwatch timing accurate to a tenth of a second is, on average, equivalent to an accuracy of about 1 kilometre in the chord measured across the minor planet.

Comets in 2004

The appearance of a bright comet is a rare event which can never be predicted in advance, because this class of object travels round the Sun in enormous orbits with periods which may well be many thousands of years. There are therefore no records of the previous appearances of these bodies, and we are unable to follow their wanderings through space.

Comets of short period, on the other hand, return at regular intervals, and attract a good deal of attention from astronomers. Unfortunately they are all faint objects, and are recovered and followed by photographic methods using large telescopes. Most of these short-period comets travel in orbits of small inclination which reach out to the orbit of Jupiter, and it is this planet that is mainly responsible for the severe perturbations that many of these comets undergo. Unlike the planets, comets may be seen in any part of the sky, but since their distances from the Earth are similar to those of the planets their apparent movements in the sky are also somewhat similar, and some of them may he followed for long periods of time.

One of the most famous periodical comets, Comet 2P/Encke, reaches perihelion (its closest approach to the Sun) on 2003 December 29, and should be visible to observers with small telescopes in December 2003 and January 2004. An ephemeris is given below:

Date 2003	2000.0 RA		Dec.		Distance from Earth A.U.	Distance from Sun A.U.	Elong- ation from Sun °	Mag.
	h	m	°	′				
Dec 1	18	22.1	+ 5	54	0.319	0.768	39.6	+ 6.8
11	17	28.4	− 7	24	0.429	0.585	16.2	+ 5.7
21	16	58.7	−17	03	0.621	0.413	14.5	+ 4.2
31	17	10.2	−24	00	0.922	0.340	20.2	+ 3.8

Date 2004		2000.0		Distance from Earth	Distance from Sun	Elong- ation from Sun	Mag.
	RA		Dec.				
	h m	°	′	A.U.	A.U.	°	
Jan 10	17 59.5	−27	24	1.231	0.448	19.5	+ 6.2
20	18 50.3	−27	54	1.472	0.626	18.7	+ 8.8
30	19 32.5	−27	04	1.666	0.808	19.5	+10.7

The following comets are expected to return to perihelion in 2004, and to be brighter than magnitude +15.

Comet	Year of Discovery	Period (years)	Predicted Date of Perihelion 2004
43P/Wolf-Harrington	1925	6.5	Mar 17
88P/Howell	1981	5.5	Apr 12
2002 T7 (LINEAR)	2002	−	Apr 23
2001 Q4 (NEAT)	2001	−	May 15
2001 HT50 (LINEAR-NEAT)	2001	−	July 9
29P/Schwassmann-Wachmann(1)	1927	14.7	July 10
121P/Shoemaker-Holt(2)	1989	8.0	Sep 1
2002 O7 (LINEAR)	2002	−	Sep 22
48P/Johnson	1963	7.0	Oct 11
78P/Gehrels 2	1973	7.2	Oct 27
69P/Taylor	1915	7.0	Nov 30
62P/Tsuchinshan(2)	1965	6.6	Dec 7

Minor Planets in 2004

Although many thousands of minor planets (asteroids) are known to exist, only a few thousand of them have well-determined orbits and are listed in the catalogues. Most of these orbits lie entirely between the orbits of Mars and Jupiter. All these bodies are quite small, and even the largest, Ceres, is only 913 kilometres (567 miles) in diameter. Thus, they are necessarily faint objects, and although a number of them are within the reach of a small telescope, few of them ever attain any considerable brightness. The first four that were discovered are named Ceres, Pallas, Juno and Vesta. Actually the largest four minor planets are Ceres, Pallas, Vesta and Hygeia. Vesta can occasionally be seen with the naked eye, and this is most likely to happen when an opposition occurs near June, since Vesta would then be at perihelion. Below are ephemerides for Ceres, Pallas and Vesta in 2004; Juno is not favourably placed for observation during the year.

1 Ceres

		2000.0 RA		Dec.		Geo-centric Distance	Helio-centric Distance	Phase Angle	Visual Magni-tude	Elong-ation
		h	m	°	′			°		°
Jan	01	7	34.14	+29	14.8	1.641	2.608	4.9	6.9	166.9W
	11	7	24.08	+30	16.7	1.625	2.602	3.1	6.8	171.6E
	21	7	13.86	+31	06.9	1.639	2.597	6.2	7.0	163.3E
	31	7	04.86	+31	42.5	1.679	2.591	10.2	7.2	152.1E
Feb	10	6	58.23	+32	03.4	1.744	2.586	13.9	7.4	141.0E
	20	6	54.64	+32	11.7	1.828	2.581	17.0	7.6	130.4E
Mar	01	6	54.36	+32	10.0	1.928	2.577	19.4	7.8	120.5E
	11	6	57.29	+32	00.5	2.039	2.572	21.1	8.0	111.3E
	21	7	03.12	+31	44.6	2.158	2.568	22.2	8.2	102.7E
	31	7	11.49	+31	23.0	2.281	2.565	22.8	8.3	94.8E
Apr	10	7	22.00	+30	55.6	2.404	2.561	23.0	8.4	87.3E
	20	7	34.24	+30	22.3	2.527	2.558	22.8	8.5	80.3E

1 Ceres (cont'd)

		RA		Dec.		Geo-centric Distance	Helio-centric Distance	Phase Angle	Visual Magni-tude	Elong-ation
		h	m	°	′			°		°
Apr	30	7	47.92	+29	42.6	2.648	2.555	22.2	8.6	73.7E
May	10	8	02.70	+28	56.3	2.764	2.553	21.4	8.7	67.5E
	20	8	18.34	+28	03.2	2.875	2.551	20.4	8.7	61.5E
	30	8	34.64	+27	03.0	2.979	2.549	19.2	8.8	55.7E
June	09	8	51.40	+25	55.8	3.076	2.548	17.8	8.8	50.2E
	19	9	8.49	+24	41.9	3.166	2.547	16.3	8.8	44.8E
	29	9	25.79	+23	21.6	3.247	2.547	14.7	8.8	39.6E
July	09	9	43.20	+21	55.3	3.319	2.546	13.1	8.8	34.5E
Nov	16	13	23.92	0	20.9	3.321	2.576	12.8	8.8	35.2W
	26	13	39.84	− 1	50.0	3.245	2.581	14.5	8.8	40.9W
Dec	06	13	55.41	− 3	12.7	3.160	2.585	16.1	8.8	46.7W
	16	14	10.57	− 4	28.3	3.065	2.591	17.6	8.8	52.8W
	26	14	25.19	− 5	36.1	2.962	2.596	19.0	8.8	59.0W

2 Pallas

		RA		Dec.		Geo-centric Distance	Helio-centric Distance	Phase Angle	Visual Magni-tude	Elong-ation
		h	m	°	′			°		°
Sept	07	9	01.47	− 3	26.3	2.924	2.136	14.4	9.0	31.9W
	17	9	21.89	− 4	24.2	2.879	2.135	15.7	9.0	35.0W
	27	9	41.85	− 5	25.3	2.827	2.135	17.0	9.0	38.5W
Oct	07	10	01.33	− 6	27.8	2.768	2.138	18.3	9.0	42.3W
	17	10	20.27	− 7	30.1	2.701	2.141	19.7	9.0	46.6W
	27	10	38.63	− 8	30.2	2.625	2.146	21.1	9.0	51.2W
Nov	06	10	56.34	− 9	26.1	2.542	2.153	22.5	9.0	56.1W
	16	11	13.32	−10	15.6	2.450	2.161	23.7	8.9	61.5W
	26	11	29.46	−10	56.2	2.352	2.170	24.8	8.9	67.2W
Dec	06	11	44.64	−11	25.2	2.247	2.181	25.7	8.8	73.4W
	16	11	58.68	−11	39.5	2.136	2.193	26.2	8.7	80.1W
	26	12	11.34	−11	35.5	2.022	2.206	26.4	8.6	87.3W

4 Vesta

		RA		Dec.		Geo-centric Distance	Helio-centric Distance	Phase Angle	Visual Magni-tude	Elong-ation
		h	m	°	′			°		°
Mar	01	20	41.57	−19	11.3	2.957	2.193	14.2	7.9	33.0W
	11	21	02.28	−18	07.8	2.896	2.199	16.1	7.9	38.0W
	21	21	22.33	−16	59.3	2.827	2.206	17.9	7.9	43.0W
	31	21	41.66	−15	47.7	2.751	2.214	19.6	7.9	48.2W
Apr	10	22	00.23	−14	34.7	2.668	2.221	21.2	7.9	53.4W
	20	22	18.02	−13	22.0	2.579	2.229	22.7	7.9	58.7W
	30	22	34.94	−12	11.3	2.485	2.238	23.9	7.8	64.2W
May	10	22	50.94	−11	04.7	2.385	2.246	25.0	7.8	69.8W
	20	23	05.93	−10	03.9	2.282	2.255	25.8	7.7	75.6W
	30	23	19.78	− 9	11.1	2.175	2.264	26.3	7.6	81.7W
Jun	09	23	32.34	− 8	28.1	2.067	2.273	26.5	7.5	88.1W
	19	23	43.43	− 7	57.3	1.958	2.283	26.3	7.4	94.9W
	29	23	52.78	− 7	40.8	1.851	2.292	25.7	7.3	102.2W
July	09	0	00.14	− 7	40.5	1.748	2.302	24.5	7.1	109.9W
	19	0	05.20	− 7	58.1	1.650	2.312	22.8	7.0	118.3W
	29	0	07.63	− 8	34.7	1.561	2.322	20.3	6.8	127.3W
Aug	08	0	07.24	− 9	29.3	1.485	2.332	17.2	6.6	137.0W
	18	0	03.95	−10	39.3	1.425	2.342	13.5	6.4	147.3W
	28	23	57.99	−11	58.7	1.386	2.352	9.4	6.3	157.6W
Sept	07	23	50.00	−13	18.9	1.370	2.362	5.7	6.1	166.4W
	17	23	40.95	−14	30.4	1.380	2.371	5.2	6.1	167.7W
	27	23	32.08	−15	24.4	1.415	2.381	8.3	6.3	159.9E
Oct	07	23	24.56	−15	55.4	1.475	2.391	12.2	6.5	149.6E
	17	23	19.24	−16	02.1	1.556	2.401	15.7	6.7	139.2E
	27	23	16.61	−15	45.6	1.656	2.410	18.6	7.0	129.2E
Nov	06	23	16.75	−15	09.0	1.769	2.420	20.8	7.2	119.8E
	16	23	19.53	−14	15.7	1.894	2.429	22.4	7.4	110.9E
	26	23	24.67	−13	08.6	2.025	2.438	23.3	7.6	102.5E
Dec	06	23	31.83	−11	50.5	2.161	2.447	23.7	7.7	94.6E
	16	23	40.69	−10	23.5	2.299	2.456	23.6	7.9	87.1E
	26	23	50.97	− 8	49.5	2.436	2.464	23.1	8.0	80.0E

Meteors in 2004

Meteors ('shooting stars') may be seen on any clear moonless night, but on certain nights of the year their number increases noticeably. This occurs when the Earth chances to intersect a concentration of meteoric dust moving in an orbit around the Sun. If the dust is well spread out in space, the resulting shower of meteors may last for several days. The word 'shower' must not be misinterpreted – only on very rare occasions have the meteors been so numerous as to resemble snowflakes falling.

If the meteor tracks are marked on a star map and traced backwards, a number of them will be found to intersect in a point (or a small area of the sky) which marks the radiant of the shower. This gives the direction from which the meteors have come.

The following table gives some of the more easily observed showers with their radiants; interference by moonlight is shown by the letter M.

Limiting Dates	Shower	Maximum	RA		Dec.	
			h	m	°	
Jan 1–6	Quadrantids	Jan 4	15	28	+50	M
April 19–25	Lyrids	Apr 22	18	08	+32	
May 1–8	Aquarids	May 4	22	20	−01	M
June 17–26	Ophiuchids	June 19	17	20	−20	
July 15–Aug 15	Delta Aquarids	July 29	22	36	−17	M
July 15–Aug 20	Piscis Australids	July 31	22	40	−30	M
July 15–Aug 20	Capricornids	Aug 2	20	36	−10	M
July 23–Aug 20	Perseids	Aug 12	3	04	+58	
Oct 16–27	Orionids	Oct 20	6	24	+15	
Oct 20–Nov 30	Taurids	Nov 3	3	44	+14	M
Nov 15–20	Leonids	Nov 17	10	08	+22	
Dec 7–16	Geminids	Dec 13	7	32	+33	
Dec 17–25	Ursids	Dec 22	14	28	+78	M

Some Events in 2005

ECLIPSES

There will be three eclipses, two of the Sun and one of the Moon.

April 8:	annular total eclipse of the Sun – the Americas.
October 3:	annular eclipse of the Sun – Africa, Europe, Asia.
October 17:	partial eclipse of the Moon – the Americas, Australasia, Asia.

THE PLANETS

Mercury may be seen more easily from northern latitudes in the evenings about the time of greatest eastern elongation (March 12) and in the mornings about the time of greatest western elongation (August 23). In the Southern Hemisphere the corresponding most favourable dates are around April 26 (mornings) and November 3 (evenings).

Venus is visible in the mornings in January. From May to the end of the year it is visible in the evenings.

Mars is at opposition on November 7 in Aries.

Jupiter is at opposition on April 3 in Virgo.

Saturn is at opposition on January 13 in Gemini.

Uranus is at opposition on September 1 in Aquarius.

Neptune is at opposition on August 8 in Capricornus.

Pluto is at opposition on June 14 in Serpens.

Part II

Article Section

Cosmic and Terrestrial Elements

PAUL MURDIN

Philosophy and science both begin when anyone asks a curious question, trying to understand why things are the way they are. The first people whom we know to have been curious in this way were the Greeks. They were the first recognizable scientists and they hit quickly on the profitable track towards understanding the structure of the scientific world. They tried to analyse what was happening in front of them in terms of simpler components, and they tried to be economical in inventing what the components were, reusing the same ones in different ways to explain different things. What we perceive about things are qualities like shape, colour, taste or change. The Greek scientists and philosophers tried to identify their underlying causes in terms of the constitution of the things perceived. We would call these underlying constituents 'elements'.

IDEAS OF THE GREEK PHILOSOPHERS

The first known Greek philosopher was Thales of Miletus (624–546 BC). None of his works survive. We know about him, and what he thought, only from the reports and criticisms by later writers. He was born in the coastal town of Miletus, in Ionia in Asia Minor (now part of Turkey). He is regarded as the founder of the Milesian school of philosophy, the first group of philosophers to raise the question of the nature of the substance from which all things are made, from which they are formed and to which they dissolve.

Miletus itself was a trading centre, a prosperous city with a great port centred in a region of fruit trees and agricultural land. Agriculture made Thales a rich man. Aristotle tells how Thales used his meteorological skills one winter to predict that the next season's olive crop would be a very large one. He therefore took options on the use of all the olive presses between Miletus and Chios and then made a fortune from his monopoly position when the great olive crop did indeed arrive.

Not all stories about Thales are favourable. Plato says that Thales was gazing at the sky one night and fell into a ditch. Seeing this, a servant girl asked him, 'How do you expect to understand what is going on up in the sky if you do not even see what is at your feet?' (This was the first absent-minded-professor joke.)

It was Thales who first thought of the scientific idea of explaining lots of phenomena by a small number of hypotheses. Specifically, he searched for an underlying reason why there was such variety in the properties of materials. He taught that 'all things are water'. It is not known why he came to this conjecture. Of course, water is found everywhere and in different circumstances, and in Egypt, which he had visited, it was associated with the surge of biological activity that accompanied the flooding of the Nile at the start of the Egyptian year. Thales, dwelling at the coast, must have observed that the Sun evaporates water, that mist rises from the ground and from the sea to form clouds, which in turn make rain. Water certainly manifests itself in different forms, and Thales evidently guessed that it was the constituent of everything – the fundamental element.

Whatever Thales's reasons for his conjecture, it was taken up, adapted and developed for 300 years. His pupil Anaximander (c.611–547 BC) could not accept that water could be both moist and dry, both cool and hot, and he thought there must be four elements – earth, water, air and fire. His theory of the Universe was that objects are formed from a vortex process by which light objects were flung out to the periphery of the vortex. This separated hot and cold, moist and dry, with Earth at the centre of the vortex, and then successive onion shells of water, air and fire. The ground under our feet, clouds and the atmosphere above our heads and the celestial bodies at great distances were evidence of this layering

In his turn, Anaximander's pupil Anaximenes (c.585–c.525 BC) took over and superseded Anaximander. He believed the primary substance of the Universe was air, which could form the other elements of water, earth and fire by compression and rarefaction. For example, water was made from dense air, which, when it vaporized, made the Sun, Moon and planets.

The idea that there are four elements was taken up by Plato (427–360 BC) and his pupil Aristotle (384–322 BC). Fire, earth, air and water were the recognizable manifestations of abstract qualities of hot, cold, dry and moist. I think that it is fair to say that with Platonism and Aristoteleanism, Greek philosophy moved away from what modern scientists would feel was science. It became too vague – playing with words rather than investigation. Even when Platonic ideas were linked to alchemists' experiments, as they tried to find a way to make gold, they put forward contradictory ideas about the elements, at one time identifying spirit, salt, sulphur, water and earth as the elements (now five of them), at another time water, oil, air, salt and earth.

The fifth element had particular cosmic significance. The seven planets (at that time Mercury, Venus, Mars, Jupiter, Saturn, the Sun and the Moon) were thought to transmit their influence to the Earth through a medium composed of a 'fifth essence'. In the Latin of the alchemists in the Middle Ages, this element was *quinta essentia*, an expression now more familiar as the term *quintessence*. It gradually evolved into the concept of the *aether*, a Greek word meaning 'upper air'. In its modernized spelling, 'ether' was the hypothetical medium that pervades the Universe, the medium that was thought in the nine-teenth century to vibrate when light waves propagated from place to place. The physicist Albert Michelson (1852–1931) tried in 1880 to detect the motion through the ether of the Earth, in its orbit round the Sun. His failure to do so was one of the motivations for Albert Einstein (1879–1955) to develop his theory of relativity. These developments disproved the existence of this special cosmic element pervading the Universe.

FROM ALCHEMY TO CHEMISTRY

In the Middle Ages, the arrangement of the planets on the zodiac, or astrology, was part of the study of alchemy, or how chemical compounds changed from one to another. The large-scale Universe, or 'macrocosm', the name of which is derived from the Greek *makrokos-mos*, 'large world', affects what happens on Earth, the 'microcosm' (*mikrokosmos* or 'small world'). The parallels between planets and elements are reflected in their old names. Quicksilver is still called mercury but similar cosmic names for other metals have fallen into dis-use. Copper was known as Venus; iron was Mars; tin was Jupiter; lead was Saturn; gold corresponded to the Sun, and silver to the Moon. It is not so long ago that lead pipes were stamped with the symbol for the planet Saturn to identify what they were made of. There was a water pipe in my house, built in 1896, which showed this, before it was taken away when the lead pipes were replaced with ones made of non-toxic materials. The main colours and metals of heraldry correspond to the colours associated with these planets: Mercury was matched with *purpure*, purple; Venus with *vert*, green; Mars with *gules*, red; Jupiter with *azure*, blue; Saturn with *sable*, black; the Sun with *or*, gold; and the Moon with *argent*, silver.

Eventually alchemy made the transition to the modern science of chemistry. Antoine Lavoisier (1743–94) and Joseph Priestley (1733–1804) experimented with making and decomposing what we now call metal oxides and identified the concept of the elements as chemicals that could not be decomposed into lighter ones. John Dalton (1766–1844) developed the idea that materials have their different properties as a result of the physical properties of the atoms of which they are made. The wooden models of atoms made by Dalton can be seen in the Science Museum in London.

The Russian chemist Dmitri Mendeleev (1834–1909) discovered an arrangement of the atoms called the Periodic Table that showed how elements were related. We now understand that the atoms of a chemical element differ from the atoms of another in the number of electrons that orbit the nucleus of the atoms. Hydrogen is the simplest element; its atoms have one electron. Iron is quite complex with twenty-eight. The Periodic Table turned out to be a list of the atoms in order of the number of electrons they had in orbit.

The arrangement of the Periodic Table originally had holes – there were locations in the Table that were empty. Naturally this sparked a search for the missing elements. The association of elements with planets continued in the way that some of the elements were named. The neurologist Oliver Sacks tells of his fascination with chemistry while he lived in London as a young boy, in his autobiography *Uncle Tungsten: memoirs of a chemical boyhood* (Picador, London, 2002, p.61).

[Some] elements had astronomical names. There was uranium, discovered in the eighteenth century and named for the planet Uranus; and a few years later, palladium and cerium, named after the recently discovered asteroids Pallas and Ceres. Tellurium had a fine earthy Greek name, and it was only natural that when its lighter analogue was found, it should be named selenium, after the moon.

Curiously, Sacks does not mention neptunium and plutonium in his list, named for the planets Neptune and Pluto. With uranium, these modern coinages complete the verbal association of all of the planets with chemical elements, as well as the moon and two asteroids.

THE APPLICATION OF SPECTROSCOPY

These associations between the elements and the cosmos were linguistic and fanciful. The first clues that the terrestrial elements were to be found in the cosmos came from spectroscopy of sunlight. In 1802 William Wollaston (1766–1828) discovered that the spectrum of sunlight had seven gaps, which he regarded as boundaries between the natural colours of the spectrum. But the optician Joseph Fraunhofer (1787–1826) used a spectroscope with superior resolution and saw, not seven, but hundreds of gaps in the solar spectrum. The gaps are now known as the 'Fraunhofer lines'. Fraunhofer accurately measured their wavelengths and labelled the stronger ones with letters.

In a crucial step forward, the chemists Robert Bunsen (1811–99) and Gustav Kirchhoff (1824–87) identified many of the Fraunhofer lines with spectral emissions made when materials such as salt were vaporized. For example, one coincidence tracked down was between the Fraunhofer D-lines in the solar spectrum and the yellow sodium emission from salt.

In 1835, in a famously inaccurate forecast, the French philosopher Auguste Comte (1798–1857) wrote of stars that, 'We understand the possibility of determining their shapes, their distances, their sizes and their movements; whereas we would never know how to study by any means their chemical composition.' By the end of the 1880s, fifty of the then known elements had been identified in the solar spectrum. This proved that the Sun was made of similar elements to the Earth. Applying spectroscopy to the brighter stars, Henry Draper (1837–82) and William Huggins (1824–1910) showed that they too had dark lines in their spectra. The spectroscopists Father Angelo Secchi (1818–78), H.C. Vogel (1841–1907) and E.C. Pickering (1846–1919) developed schemes for classifying the spectra of stars and listing the Fraunhofer and other spectral lines that were found in each. The same elements had been found, not only on the Earth and the Sun, but also in the stars and indeed the nebulae too.

In 1868–9 there was a dramatic development. Norman Lockyer (1836–1920) and Jules Janssen (1824–1907) observed the solar chromosphere during the total solar eclipse of 1868 and saw and measured a strong spectral emission at a wavelength near to the sodium D-lines. Janssen realized that it would be possible to view the spectral lines of the chromosphere even without an eclipse being in progress. In the calm atmosphere of scientific study that a solar observatory affords, more relaxed than during a brief eclipse, the wavelength of the chromospheric emission could be accurately determined and it proved to be different from the sodium D-lines, indeed different from all known emissions from known elements.

Janssen and Lockyer both realized that they might have discovered a previously unknown element. Lockyer named it helium, after the Greek for the Sun, *helios*. This element was isolated on Earth in 1895 by the Scots chemist William Ramsay (1852–1916), in radioactive minerals where it is formed as a decay product. Thus helium was discovered as an element in a cosmic object before its identification on Earth.

I visited University College once, in Gower Street, London, and was shown into what is now the Slade School of Art to see the place where Ramsay's discovery was made. It is commemorated by a plaque, which could, at the time I was there, be read only by me and the naked model in front of a life drawing class, the members of which had their backs to the wall. I hurriedly read the plaque, made an excuse and, averting my eyes, quickly left.

Further new spectral lines were discovered in the chromospheric spectrum of the total solar eclipse of 1869. The new element coronium that was invented to account for these spectral lines turned out not to be real. The astronomer Walter Grottrian (1890–1954) showed theoretically in 1941 that coronium is in reality iron at high temperatures and low densities. Similarly, spectral lines in nebulae were once attributed to the element nebulium, a misidentification of oxygen and other common elements, for very similar reasons. These are just two of many imagined elements, proposed on spurious theoretical grounds or misidentified both in stellar spectra and on the Earth. Sacks lists several:

> In addition to the hundred-odd names of existing elements, there was at least twice that number for elements that never made it, elements imagined or claimed to exist on the basis of unique chemical

or spectroscopic characteristics, but later found to be known elements or mixtures . . .

I was oddly moved by these fictional elements and their names, especially the starry ones. The most beautiful, to my ears, were aldebarium, and cassiopeium (Auer's names for elements that actually existed, ytterbium and lutecium) and denebium, for a mythical rare earth. There had been a cosmium and neutronium (element 0) too . . .

Aldebarium and denebium were named after the stars Aldebaran and Deneb respectively, and cassiopeium after the constellation Cassiopeia. The latter name is still used in Germany for the element lutecium.

THE MODERN VIEW

Whatever the false turns to the path by which history has brought us to the present day, all one hundred spaces in Mendeleev's Periodic Table are now filled in with valid identifications of chemical elements, nearly all now also identified in the solar spectrum and many in the stars. In fact, the chemical elements are a thread that ties together our knowledge of the evolution of stars, galaxies and the Universe. The origin of the Universe in the Big Bang yielded the simplest and most abundant of the elements, hydrogen and helium. Most other elements have been produced by the nuclear reactions that power the stars during the normal course of their evolution, or during explosions of certain stars (like supernovae). Indeed, we now understand that the elements were carried in the interstellar medium from their cosmic origins into the solar nebula from which the Sun formed and into the planets, including our own. The Earth and all that is in it, including us, are made of cosmic elements. We are a part of the Universe, not apart from it. The verbal distinction between the cosmic elements and the terrestrial ones, as in the title of this article, makes little scientific sense.

We now know all the chemical elements, but this is not of course the conclusion of the process of discovery of the constituents of the Universe. We are amazingly ignorant about its fundamental contents. We do at least know how ignorant we are. According to the classification scheme of US Secretary of Defense Donald H. Rumsfeld (BBC

Today programme Rumsfeld soundbite of the year 2002): 'There are known knowns; things we know that we know. There are known unknowns: things we know we don't know. But there are unknown unknowns: things we do not know we don't know. Each year we discover further unknown unknowns.' Astronomers have discovered two unknown unknowns and moved them into the category of known unknowns by mapping the distribution of matter in the Universe and the way it is moving, expanding outwards. They infer that the chemical elements are no more than 5 per cent of the contents of the Universe. Some 25 per cent of the Universe is made up of 'dark matter' and 70 per cent of the Universe is made up of a form of 'dark energy'.

Dark matter might be some form of exotic particle, and there are experiments on the Earth, for example in the Boulby Mine in Yorkshire, trying to detect these particles here on Earth. Dark energy is conceived as a property of space. Particle physicists have developed a theory of dark energy called, in a playful mood, 'quintessence', referring back to its equivalent of the Middle Ages. Dark energy or quintessense is available in the space here on Earth. In principle the energy might even be released by human intervention, with enormous practical benefit. If dark matter and dark energy are found by laboratory experiments they will be further cases where fundamental constituents of matter, like the element helium, have been discovered in the cosmos before being identified on Earth. This is for the future to show, but if they happen, the discoveries will be part of a scientific process whose origins are at least three thousand years old.

Cassini–Huygens to Saturn

DAVID M. HARLAND

Within a few years of the invention of the telescope in the early seventeenth century, Saturn was found to possess a satellite, which was named Titan, and, seemingly uniquely among the planets, to be surrounded by a ring. By the mid nineteenth century it had become evident that this feature comprised a series of concentric rings. Until the discovery of Uranus by William Herschel in 1781, Saturn was believed to be the outermost of the planets, orbiting the Sun at a mean distance of 1.4 billion kilometres with a period of almost thirty years. Like Jupiter, the largest member of the solar retinue, Saturn is a hydrogen gas giant, but with only one-third of its mass. While the surface of its atmosphere shows similar latitudinal banding to that of Jupiter, it is much less striking. Although Saturn's equatorial diameter is 120,000 kilometres, its polar diameter is some 10 per cent less, making it the most oblate of the planets. By the dawn of the 'Space Age', our knowledge of Saturn's system of moons focused on the 'resonances' between their orbital periods; almost nothing was known of their physical form – with the exception of Titan, they were mere specks of light in even the largest telescopes.

FIRST SPACECRAFT STUDIES

The first spacecraft to reach Saturn was Pioneer 11 in September 1979. It served as a pathfinder for the Voyagers which passed through the system in November 1980 and August 1981. Most of what we know of Saturn, its moons and its magnetosphere derive from these missions. In terms of their sizes, and working outwards from the planet, Saturn's classically named moons can be grouped with Mimas and Enceladus at about 400 to 500 kilometres in diameter, Tethys and Dione at about 1,000 kilometres, and Rhea and Iapetus at 1,500 kilometres. Studies of orbital resonances implied densities suggesting that they were primarily icy bodies. The ring system lies in the plane of the planet's equator and, in terms of planetocentric distance (which is a convenient scale for measuring distances in a planetary system) its most prominent part extends out to about 2.3 radii. Mimas travels in a more or less circular orbit at 3.1 radii, in the same plane, with a period of 22.6 hours.

The Voyagers showed Mimas to be intensely cratered, with one crater having a diameter fully one-third that of the moon – it is remarkable that the moon survived the impact. Although Enceladus had been expected to be similar to Mimas, it proved to be a surprise. Much of its surface is smooth, indicating that it has been resurfaced by cryovolcanic activity – perhaps recently. Of the two medium-sized moons, Tethys has been heavily battered and has a crater into which Mimas could snuggly be fitted, together with a world-girdling system of canyons whose formation may have occurred at the same time as this impact. Dione is cratered too, but its most remarkable feature turned out to be dark mottling on its trailing hemisphere, upon which is superimposed a pattern of bright wispy streaks whose origin is a mystery. There is a similar albedo feature on Rhea, the next moon out.

Titan, the largest of Saturn's moons, lies at about 22 radii, orbiting with a period of sixteen days. At 5,150 kilometres in diameter, it is only marginally smaller than the Jovian moon Ganymede, which is not only the largest planetary satellite in the Solar System but is also larger than the planet Mercury. Titan is unique among the moons in possessing a dense atmosphere. Spectrographic observations in 1944 identified methane in the atmosphere, but it was not known whether this was the main constituent. When an image by Pioneer 11 showed Titan's atmosphere to be optically thick, it was realized to be dense. Remote sensing by Voyager 1 established that it was primarily composed of nitrogen. Following the break-up of methane in the upper atmosphere by ultra-violet insolation, polymerization has formed a variety of aerosols that form layers of haze.

Although the surface pressure of 1.5 bars was comparable to that at the surface of the Earth, this was a remarkably high value for such a small body as Titan on which gravity is much weaker; in terms of the sheer bulk of gas, Titan's atmosphere is ten times denser than ours. The surface temperature is 90 K. It may rain ethane and there may be large bodies of liquid hydrocarbons in surface depressions, and much of the surface may be coated with a thick hydrocarbon sludge. Overall, conditions on Titan may resemble a cryogenic version of the early

Earth's atmosphere, and may be in a 'pre-biotic' state. Hyperion, orbiting beyond Titan, was found to be an irregular ($360 \times 280 \times 236$ kilometres) body that appears to be a fragment of a much larger object. Iapetus, far beyond Titan, and in an inclined orbit, had long posed a mystery by virtue of having a distinctly darker leading hemisphere. The Voyagers found this line of dichotomy to be well defined, and the dark region to be essentially featureless. Even further out is Phoebe. It traces an eccentric orbit which is highly inclined to the plane of the Saturnian system but is close to the plane of the ecliptic and, as such, is likely to be a captured asteroid or cometary nucleus – although at 200 kilometres in diameter it is a very large one.

THE CASSINI–HUYGENS MISSION

The Pioneer and Voyager fly-bys provided only snapshots. In 1982 it was decided to send a spacecraft to orbit Saturn to undertake an in-depth investigation of the system, to map the surface of Titan by radar and to drop a probe into its atmosphere. This Cassini–Huygens mission is a collaboration between America and Europe, with NASA supplying the Cassini spacecraft, the European Space Agency supplying the Huygens probe and the Italian Space Agency supplying the communications system for relaying data from the probe to the main spacecraft. Development was slow, essentially due to financial constraints, and the mission was 'de-scoped' several times before the final configuration was agreed in 1992 and launch set for 1997. The Cassini spacecraft is the most sophisticated vehicle ever dispatched into the outer Solar System. In addition to carrying and delivering the Huygens probe to Titan, it carries 335 kilograms of instruments designed to facilitate comprehensive and complementary observations of the particles and fields both in interplanetary space and in planetary magnetospheres, and multispectral remote sensing of Saturn, its rings and moons. The joint Cassini–Huygens spacecraft has eighteen instruments, twelve on the orbiter and six on the Huygens probe, many of which are capable of multiple functions (see Table 1, p. 172).

Cassini's primary imaging instrument is a CCD camera system with independent wide- and narrow-angle fields of view sensitive from the ultraviolet through the visual range into the near-infrared, with a variety of filters to facilitate specific types of observation. This is

Figure 1. The Huygens probe is installed on the Cassini orbiter in the Payload Hazardous Servicing Facility at the Kennedy Space Center in July 1997. (Image courtesy of NASA/JPL.)

supplemented by a visual and infrared mapping spectrometer, a composite infrared spectrometer and an ultraviolet imaging spectrometer. The four-metre-diameter high-gain communications dish antenna will double up as a microwave radar to map the surface of Titan through its obscuring atmosphere. Once the 2,150-kilogram spacecraft was filled with 3,132 kilograms of propellant and loaded with the 373-kilogram Huygens probe, it was the heaviest spacecraft yet sent into deep space.

Launched on October 15, 1997 by a Titan IV-B/Centaur, Cassini began its long journey to Saturn by heading inside the Earth's orbit

Figure 2. Blast off! The Cassini–Huygens spacecraft begins its seven-year journey to Saturn. The successful launch, aboard a Titan IV-B/Centaur, occurred at 9:43 UT on October 15, 1997. (Image courtesy of NASA/JPL.)

for a fly-by with Venus on April 26, 1998, where it barely skimmed the planet's atmosphere. Operational limitations restricted observations. The gravitational 'slingshot' eased the spacecraft's trajectory out beyond the Earth's orbit and on December 3, four days before the 1.58 AU aphelion, the main engine was fired for eighty-eight minutes to set up a second encounter with Venus. On its way out from the encounter of June 24, 1999, Cassini flew by the Earth at an altitude of 1,166 kilometres on August 18. The Earth pass was the first opportunity to check the calibrations of some of the spacecraft's instruments. On January 23, 2000, Cassini made multispectral observations of 2685 Masursky (an asteroid named after the renowned planetary geologist Hal Masursky, who died in 1990). Although at a range of 1.6 million kilometres the asteroid spanned no more than a few pixels, this brief encounter was a welcome test of the mutual boresighting of the battery of remote-sensing instruments.

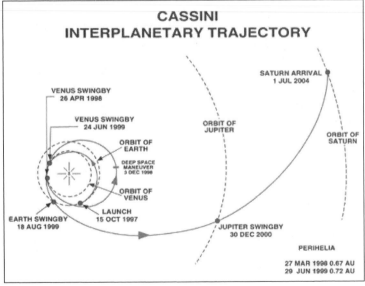

Figure 3. Interplanetary flight path of Cassini–Huygens, beginning with launch from Earth on October 15, 1997, followed by gravity-assisted fly-bys of Venus (April 26, 1998 and June 24, 1999), Earth (August 18, 1999), and Jupiter (December 30, 2000). Arrival at Saturn is scheduled for July 1, 2004, which marks the beginning of a four-year orbital tour of the Saturn system. (Image courtesy of NASA/JPL.)

JUPITER FLY-BY

Cassini's passage through Jovian space towards the end of 2000 was a full dress rehearsal for the science teams. Cassini made co-ordinated observations with the Galileo spacecraft that had been in orbit there since 1995, with one spacecraft reporting on the solar wind heading in towards the planet and the other one reporting upon its effects from a vantage point within the magnetosphere. As Cassini approached, it had an uninterrupted view of Jupiter's illuminated hemisphere. This provided a welcome opportunity to undertake one of the tasks that Galileo had been unable to perform as a result of the low transmission rate imposed by the failure to deploy its high-gain antenna. On October 1, 2000, when still 84 million kilometres out, Cassini started taking narrow-angle pictures to form a movie documenting 168 planetary rotations.

Figure 4. Image of Jupiter, acquired by the Cassini spacecraft from a distance of 77.6 million kilometres (48.2 million miles) on October 8, 2000. The cloud patterns reflect different physical conditions – updraughts and downdraughts – in which the clouds form. The Great Red Spot (below and to the right of centre) is a giant atmospheric storm as wide as two Earths and over 300 years old. (Image courtesy NASA/JPL and University of Arizona.)

An analysis of the atmospheric features towards the poles, where the weather patterns are less organized than at lower latitudes, turned up a major surprise. At first sight, the mottling in the polar regions appeared to be chaotic, but this was not the case. Thousands of 'small' spots (each of which was actually an active storm system larger than the largest terrestrial storm) were seen jostling one another as they streamed together in a given latitudinal band, only a few changing bands. Although a few spots merged, most persisted throughout the sequence. Terrestrial storm systems last no more than a week or so before they break up. Perhaps, instead of wondering why the Jovian storms persist so long, we ought to be asking why our own weather system is so dynamic.

With its Magnetospheric Imaging Instrument, Cassini was the first spacecraft equipped to *image* the bubble of charged particles trapped in a magnetosphere. Furthermore, because its high-gain antenna could operate 'passively' as a radiometer, it was able to detect the 'synchrotron' emission from the electrons which were spiralling along the magnetic field lines. Since the magnetic axis is inclined to the equatorial axis of the system, when these data were processed to form a movie, this clearly showed the precession of the radiation belt that lies in the magnetic equator. The Magnetospheric Imaging Instrument also made the discovery of a torus of material occupying the orbit of Europa, an intriguing moon which may contain a vast ocean beneath its thin icy shell. Although Europa takes three and a half days to orbit Jupiter, the inner part of the magnetosphere rotates with the planet in about ten hours. The rotation of the moon is synchronized with its orbital motion, so its trailing hemisphere is bombarded by the energetic charged particles that are trapped in the magnetosphere. This continuous rain of particles, many of which are 'heavy' ions that have been accelerated by the magnetic field, dissociate the water-ice molecules exposed at the surface, 'sputtering' them into space as a torus of neutral gas with a mass comparable to that of Io's torus (formed of material blasted into space by that moon's volcanic plumes).

Since Cassini approached Jupiter no closer than 10 million kilometres, it was not able to make high-resolution studies of the moons, but it did observe the tremendous volcanic plume from the eruption at Tvashtar on Io. Although the observations had to be curtailed due to a temporary problem with a gyroscope, a few multispectral images of Himalia were also secured. Himalia is the leading member of one of several groups of small moonlets in distant elliptical, steeply inclined orbits, and as such likely represents an asteroid that fragmented after being captured by Jupiter.

As Cassini left Jupiter behind, it took star sightings to check its trajectory for Saturn. The images of the star Spica revealed that the view through its narrow-angle camera had become 'fogged'. The cause of this impairment may never be determined, but it may have been a coating of efflux from the thrusters that were used while the gyroscopic system was out of commission. The situation was significantly improved by boosting the heater unit inside the camera, however, and as this article was being written, the camera was giving striking long-range views of Saturn.

ARRIVAL AT SATURN

On June 11, 2004, on its way into the Saturnian system, Cassini will pass within a few thousand kilometres of Phoebe. This should yield some spectacular pictures. On July 1 it will gain an early view of Enceladus from a range of 25,000 kilometres, the enigmatic moon that appears to have been 'resurfaced'. Later in the day, the spacecraft will cross the plane of the ring system at a planetocentric distance of 2.63 radii, just beyond the very narrow 'F' ring, and then close in to 1.3 radii, at which time it will fire its engine for ninety-six minutes to slow its speed sufficiently for the planet to capture it. The initial orbit will be highly eccentric, and inclined at 17° to the system's equatorial plane. The engine will be fired again at apoapsis, on August 23, this time to accelerate the spacecraft and raise the low point of its orbit away from the planet and to set up a 1,200 kilometre fly-by of Titan on its way back in, on October 26. As it passes the moon, the spacecraft's sensors will determine the prevailing winds in order to refine the plan for dropping the Huygens probe into the moon's atmosphere. The winds will be verified during the second fly-by on December 13.

After refining its trajectory on December 17 for a collision course with Titan, Cassini will release the Huygens probe on December 24. Five days later, Cassini will fire its engine so as to pass 60,000 kilometres 'in trail' of the moon. It had been intended to make a much closer fly-by on the leading hemisphere while receiving the data during

Figure 5. An artist's impression of Cassini–Huygens during the Saturn Orbit Insertion (SOI) manoeuvre, just after the main engine has begun firing. The 96-minute SOI manoeuvre will allow Cassini to be captured by Saturn's gravity into a five-month orbit. Cassini's close proximity to the planet following SOI offers a unique opportunity to observe Saturn and its rings at very high resolution. (Image courtesy of NASA/JPL.)

the probe's descent into the moon's atmosphere. However, when a flaw was found in the design of the receiver which would cause the 'received frequency' of the probe's transmission to be Doppler-shifted by the high-speed pass beyond the receiver's ability to receive it, the fly-by range had to be opened in order to reduce the Doppler shift and so maintain the received frequency within the narrower-than-planned frequency 'window' of the receiver. Furthermore, to preserve the main spacecraft's subsequent orbital tour, the fly-by had also to be reassigned to the trailing hemisphere.

DESCENT TO TITAN'S SURFACE

The Huygens probe will reach Titan on January 14, 2005. It must dive into the atmosphere at a depressed angle of 65° to establish the desired deceleration profile. The aim point is at about 11° south of the equator and longitude 190° west, near the centre of the hemisphere that permanently faces away from the planet. Due to uncertainties as to the winds, the nominal ellipse is 1,200 kilometres long and 200 kilometres wide, but if the winds can be determined prior to the dispatch of the probe this should be able to be considerably constrained, so that a particular landing site might be selected. Titan's atmosphere is so deep that the probe, travelling at just over 6 kilometres per second, will encounter it at an altitude of 1,250 kilometres. The temperature of the shock wave immediately in front of the probe's conical nose will soar to 12,000°C, and the deceleration profile will peak at 16 g.

At an altitude of 170 kilometres, well within the orangey optically thick haze, a mortar will deploy a 2.6-metre-diameter drogue parachute that will take away the 'back cover' to permit the 8.3-metre-diameter main braking chute to deploy. About thirty seconds later, the 'front shield' will be jettisoned so that the probe's instruments can begin to directly sample the atmosphere. Fifteen minutes later, the large chute will be released and replaced by a 3-metre-diameter chute designed to enable the probe to descend through the coldest layer of the atmosphere

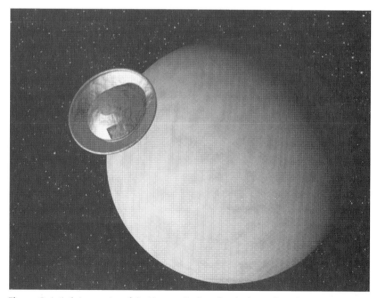

Figure 6. Artist's impression of the Huygens Probe, after deploying from the Cassini Orbiter, en route into the murky atmosphere of Saturn's largest moon, Titan. (Image courtesy of NASA/JPL.)

sufficiently rapidly to prevent its systems from freezing, and yet sufficiently slowly to hit the surface at a speed of no more than six metres per second. At about 90 K, the ice will be as hard as steel. The probe has been designed to float in case it should splash down. During its descent, it will profile temperature, pressure, density and composition in terms of altitude.

Throughout the parachute descent, as the probe is carried by the prevailing wind, an imager will take panoramic pictures to survey the ground track and to provide the context in which to interpret the results of the surface-science package. The probe's batteries should sustain three hours of full operations, which should be sufficient to reach the surface and allow a few minutes of reporting of surface conditions. The probe has no data storage device. It must transmit its data 'live'. It has two transmitters. In addition to providing a degree of redundancy, a duo of transmitters will enable one signal to be delayed by several seconds so that data lost in momentary 'drop outs' on one channel may well be able to be recovered from the other. The signal will be too weak to be received directly from Earth. The Cassini spacecraft will store it in its large solid-state memory. Cassini will maximize the strength of the received signal from the probe by slowly turning to maintain its large antenna towards the probe's path (as calculated from remote sensing of the prevailing winds during the previous fly-bys). Consequently, it will not be able to relay the data to Earth in real time. We will have to wait until Cassini has secured the entire transmission, and reorients itself to point its dish towards the Earth to replay the data from its memory. By the time we receive the first data, the probe's mission will be over.

While the scientists will investigate in detail the temperature and chemical profiles of the atmosphere and the physical properties of the surface, the general public is more likely to be captivated by the imagery providing the first view of this alien landscape, hopefully from the surface. Although there may be striking large-scale tectonic features and a great many craters, much of Titan's icy surface is likely to be coated with a black tarry goo comprising a concoction of hydrocarbons that Carl Sagan named 'tholin' – with lakes of ethane in some of the depressions. One thing is certain, it will be a very long time indeed before any human stands on Titan's surface to witness the scene in person.

CASSINI'S ORBITAL TOUR

After it has replayed the probe's data, Cassini will be free to perform its primary mission more or less as planned. If the probe cannot be released on schedule, a second opportunity is available on or about January 25, 2005, with the atmospheric entry on February 15, but if this is necessary it will significantly disrupt the rest of the mission.

As the largest moon, Titan will form the focus of the four-year primary mission, with forty-four of the sixty orbits providing close encounters during which the spacecraft will barely skim its uppermost layer of haze. On each pass, the high-gain antenna, operating as a microwave radar, will penetrate the optically thick haze to map a swath of the surface. In much the same manner as Galileo used the slingshots of close encounters with Jupiter's large moons to set up the orbital tour of that system, Cassini will use Titan encounters to progressively alter the apoapsis, orientation and inclination of its orbit around Saturn so that, over the duration of its time in orbit, it will be able to achieve a variety of objectives. Successive fly-bys will adjust the spacecraft's orbit away from the equatorial plane until at the conclusion of the primary mission it is inclined at 84° in order to facilitate a study of the circulation of the planet's atmosphere at high latitudes, to study the polar regions of the magnetosphere, and to provide strikingly 'open' views of the distribution of the material in the ring system.

Of the icy moons, Enceladus has the highest priority, having been assigned three fly-bys within 500 kilometres, phased to facilitate complete mapping. It might still be 'active', with geysers venting water vapour into space to crystallize and replenish the 'E' ring. Dione, Rhea and Iapetus are each to receive a single close inspection. Being in an inclined orbit far out beyond Titan, Iapetus will be difficult to reach, but at the right moment, as the plane of the spacecraft's orbit is progressively inclined, a close Titan fly-by will temporarily draw out the spacecraft's apoapsis sufficiently to reach the mysterious outer moon.

The Cassini–Huygens mission is facilitating twenty-seven specific scientific investigations and nine interdisciplinary studies drawing data from two or more instruments, involving Saturn's interior, atmosphere, magnetosphere, rings and moons.

If Cassini is still healthy at the end of its primary mission in 2008, its programme will undoubtedly be extended.

Table 1. Cassini–Huygens Spacecraft Instruments

Orbiter instruments:

Cassini Plasma Spectrometer (CAPS)

A direct sensing instrument to measure the energy and electrical charge of particles such as electrons and protons in the spacecraft's immediate vicinity.

Cosmic Dust Analyser (CDA)

A direct sensing instrument to measure the size, speed and direction of tiny dust grains in the spacecraft's immediate vicinity.

Composite Infrared Spectrometer (CIRS)

A remote sensing instrument to measure the infrared from an object such as an atmosphere or moon surface to determine its temperature and chemical composition.

Ion and Neutral Mass Spectrometer (INMS)

A direct sensing instrument to analyse the charged particles (electrons, protons and heavier ions) and neutral atoms in the immediate vicinity of the spacecraft.

Imaging Science Subsystem (ISS)

A remote sensing instrument to capture images across the visible wavelength range and into the infrared and ultraviolet.

Dual-technique Magnetometer (MAG)

A direct sensing instrument to measure the strength and direction of the magnetic field in the spacecraft's immediate vicinity.

Magnetospheric Imaging Instrument (MIMI)

A composite package to study the particles trapped in either the solar wind or a planetary magnetosphere. Its Low-Energy Magnetospheric Measuring System and Charge-Energy Mass Spectrometer operate by direct sensing. The Ion and Neutral Camera operates by remote sensing and (for the first time on any spacecraft) provides images.

Cassini RADAR

In addition to serving as the primary communications link, the spacecraft's main dish antenna can be used both as a passive radiometer and as an active microwave radar.

Radio and Plasma Wave Science Instrument (RPWS)

A combination package for direct and remote sensing of the electric and magnetic wave fields in the interplanetary medium and planetary magnetospheres.

Radio Science Subsystem (RSS)

Analysis of the Doppler shift on the spacecraft's radio signal during close fly-bys can yield information on gravitational fields, and the way in which the signal is refracted during occultations can yield information on the atmosphere of a planet or moon.

Ultraviolet Imaging Spectrograph (UVIS)

A remote sensing instrument to determine the way in which the atmosphere of a planet or moon reflects ultraviolet light, in order to determine its composition.

Visible and Infrared Mapping Spectrometer (VIMS)

A combination remote sensing instrument to measure visible and infrared wavelengths in order to determine the composition of a planetary atmosphere or the surface of a moon or Saturn's ring system.

Huygens probe instruments:

Huygens Atmospheric Structure Instrument (HASI)

A suite of sensors that will measure the physical and electrical properties of Titan's atmosphere.

Doppler Wind Experiment (DWE)

An analysis of the Doppler shift on the probe's transmission will provide information on the winds in Titan's atmosphere.

Descent Imager/Spectral Radiometer (DISR)

A variety of sensors will be deployed to perform a range of imaging and spectral observations, including measuring the upward and downward flow of radiation in order to determine the radiation balance of the thick atmosphere and scattering of sunlight by aerosols in the atmosphere. Visible and infrared imagers will observe the surface during the latter stages of the descent and, as the probe slowly rotates on its parachute, provide a series of panoramic views of the landing site.

Gas Chromatograph Mass Spectrometer (GCMS)

A versatile gas chemical analyser designed to identify and measure the abundance of chemicals in Titan's atmosphere.

Aerosol Collector and Pyrolyser (ACP)

It will draw in a sample of the atmosphere and decompose the complex organic materials for analysis by the GCMS.

Surface-Science Package (SSP)

A number of sensors will determine the physical properties of Titan's surface at the point of impact, whether the surface is solid or liquid.

Hypernovae

CHRIS KITCHIN

INTRODUCTION

Until less than a decade ago supernova explosions were thought to be the most intense explosions occurring in the Universe since the original Big Bang. But in February 1997 a crucial observation revealed explosions potentially a hundred times more powerful than the most energetic supernova. That observation by the BeppoSAX spacecraft was of a gamma-ray burster, or GRB (see box, p. 181, for more details), just one of many that this experiment had found. However, in this case an optical afterglow followed the gamma rays and that glow was observed spectroscopically by the Keck II telescope. The optical observations allowed a red shift for the burster to be found, and so its distance could be determined as about 9,000 million light years. Observers of GRBs had already begun to suspect on the basis of the uniform distribution of the bursts in the sky that they were to be found towards the edges of the Universe, but this was the first real proof that the idea was correct. The implications of GRBs being at immense distances were, however, that they must also be enormously powerful, emitting up to 10^{47} J in a few seconds in the brightest examples. This compares with 10^{45} J for the total energy, including the neutrinos, emitted by the brightest supernovae.

Now the release of 10^{47} J of gamma rays requires the total conversion into energy of some 10^{30} kg of matter – about half the mass of the Sun. Theories about how this might be accomplished abounded in the 1990s. Most have fallen by the wayside as observations showed that they were untenable, so that the two current leading contenders to explain GRBs are the coalescence of two neutron stars, and hypernovae.[1] Evidence is accumulating in favour of the latter at the time of writing, but it may be that both models are correct, since there seem to be two types of GRB – those with the longer bursts at lower energies perhaps coming from hypernovae, while the shorter, higher-energy bursts may arise from neutron star collisions.

What, then, are hypernovae? Essentially they are failed supernovae, and so to understand them we must first look at what is thought to occur during supernova explosions.

SUPERNOVAE

Although supernovae are now classified into a number of different types on the basis of their spectra and light curves, the two main groupings arise from population I and population II stars. Confusingly, though, it is type II supernovae that come from population I stars and type I from population II stars!

Type I supernovae need not concern us much here since the processes thought to be producing them are unrelated to those for hypernovae. They are the brightest supernovae in the optical region, sometimes reaching an absolute magnitude of -19^m (comparable with the integrated magnitude of a medium-sized galaxy). Various processes are suggested for their origins, including the complete thermonuclear explosion of the degenerate core of a highly evolved star, the cataclysmic end for a white dwarf within a close binary as accretion takes the white dwarf mass above the Chandrasekhar limit, and the result of the coalescence of two white dwarfs in a close binary system

Type II supernovae arise as stars of eight or more solar masses reach the ends of their lives. The Sun has a lifetime in the region of 10^{10} years, but although more massive stars have more material within them so that it might appear that they would outlive the Sun, they use up that material at a far faster rate and so have comparatively short lives. An eight solar mass star might thus last for a hundred million years while a thirty solar mass star will be gone in just ten million years. During their lives these stars will initially produce their energy by converting hydrogen to helium just as the Sun does at the present time. The helium will accumulate at the star's centre until the star's core becomes almost pure helium. The hydrogen 'burning' will then cease at the centre of the star, but continue in a shell around the helium core where hydrogen is still to be found. The helium core will slowly shrink and its temperature will rise until at about 100 million degrees three helium nuclei can combine to produce a single carbon nucleus. This triple-alpha reaction (helium nuclei are the alpha particles of radio-activity) will supply the star with energy for perhaps half a million years. As the temperatures and densities rise further, helium and carbon combine to form oxygen, neon, magnesium and silicon.

Once the helium has been consumed, the star's core will again shrink. Eventually, at about a million times the density of water, and at several hundred million degrees, the carbon nuclei react with each other, producing magnesium, and at somewhat higher temperatures and densities, neon, oxygen, sulphur and finally silicon also unite. The carbon 'burning' period lasts for less than a thousand years, neon and oxygen for less than a year, and the silicon burning is over in a day. Since several of these reactions may be continuing in layers outside the core, the central region of the star then resembles an onion in its struc-ture (see Figure 1). The product of the silicon burning is largely nickel-56. This is a radioactive isotope of nickel which decays to cobalt-56, and that in turn decays to iron-56. All three of these nuclei, however, are among the most unreactive known. When they are involved in nuclear reactions, energy has to be supplied from outside either to build them up to heavier elements or to break them down to lighter elements. Thus at the centre of the star the formation of the iron core (strictly a nickel core to begin with) marks the end of the road for nuclear reactions that generate energy for the star. The core shrinks, until at a density of about 10^{11} kg m^{-3} it becomes electron-degenerate. But electron-degenerate material is the stuff of white dwarfs, and as is well known there is an upper limit on the mass of a white dwarf of about 1.4 solar masses. Inside our large star the nuclear reactions in the higher layers are continually adding material to the iron core. The core thus soon exceeds the Chandrasekhar limit and collapses down in a fraction of a second to become a neutron star.

The outer layers of the star, still containing all the elements from hydrogen to iron, collapse in their turn and collide with the surface of the neutron star at a speed around 10 per cent that of light. Huge amounts of energy are released in the outer layers of the star as the light elements are consumed, and shock waves cause those outer layers to explode outwards. Vast quantities of neutrinos are also produced, and these aid the explosion by further heating the material outside the core. Thus the outer layers of the star are blasted off into space, and we see the start of a supernova. Some 10^{42} J will be emitted by the supernova in the

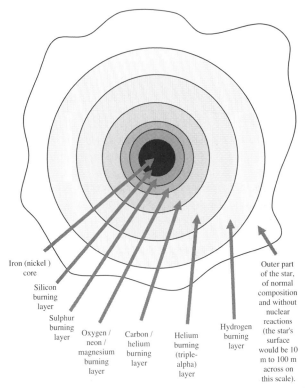

Figure 1. The structure at the centre of a massive star immediately prior to core collapse.

form of light and other electromagnetic waves, but in addition some 10^{43} J will go into the kinetic energy of the exploding material and an incredible 10^{45} J comes off in the form of neutrinos. That last figure is the total energy emitted by ten solar-type stars over their entire lifetimes – but even so it is still only 1 per cent of the energy required for a GRB.

HYPERNOVAE

If supernovae can produce up to 10^{45} J and arise from stars with masses from eight to thirty times that of the Sun, then it would seem reasonable to expect that even larger stars might produce even more energy. However, that does not appear to happen. Instead, if we follow through the collapse of, say, a forty solar mass star, the sequence of events follows that for a supernova until we get to the collapse of the core to a neutron star. Then things change in two ways. Firstly the outer layers of the large star contain so much more material than is present in the smaller stars that the shock waves and neutrinos are unable to eject the overlying material to produce a visible supernova. At most, the explosion fizzles out and produces a dim imitation of a normal supernova. The second difference from smaller stars is that the collapsing core of the huge star does not halt at the neutron star stage but collapses further to become a black hole. It may do this in an unbroken manner, becoming a black hole in less than a second. Alternatively, the neutron star may form briefly (a few seconds to an hour or so) before the addition of material from the still collapsing outer layers of the star at a rate of up to one solar mass every ten seconds takes it over the maximum mass of around three solar masses for a neutron star and it then finishes its collapse to a black hole. Either way, much of the matter and energy is trapped inside the black hole and no more is released into the outer universe than with a conventional supernova.

Thus explaining GRBs as supernovae occurring in very large stars looks to be a non-starter. However, in 1993, Stan Woosley of the University of California, Santa Cruz, realized that there was still a way in which GRBs might result from the collapse of a massive star. This was if the gamma rays were not emitted isotropically, but in the form of beams or jets of radiation and material. This would have the effect of reducing the total energy required by a factor of about a hundred, to approximately the energy resulting from a supernova. However, it would also mean that we would only observe a GRB when the jet was directed towards the Earth, and so the number of GRBs actually occurring must be about a hundred times the number that we actually see. Woosley, with his graduate student Andrew MacFadyen, developed the model over the next few years until it is now one of the main contenders to explain GRBs. The crucial factor in this success was Woosley's realization of the vital rôle played by rotation during the collapse of a supermassive star. This both enables the production of beamed radiation and allows it to escape into the outer universe. The object that results when rotation is included in the model for a collapsing star is what we now call a hypernova.

The importance of rotation is that in the equatorial plane the material falling in towards the central black hole first forms a dense accretion disc, while the polar regions are cleared as the material falls through to the black hole. Material falling into the accretion disc heats it to perhaps 2×10^{10} degrees, and in the ensuing chaos, myriads of subatomic particles including electrons, protons, photons, neutrinos and their antiparticles are produced, all moving at relativistic velocities. Only the neutrinos, though, can escape quickly from the disc and they interact to produce electrons, positrons and photons around the black hole. These in turn try to expand outwards, but can only do so through the comparatively clear polar regions. The gamma rays that we observe as a GRB are then produced from interactions within the jets and as the jets impact into material away from the core. The energy injected into the jets and so into the gamma rays is estimated to be up to 10^{45} J – comparable with the total energy from a supernova, and sufficient now to power a GRB because it is only going into the two beams that cover about 1 per cent of the sky.

There are, of course, still competing models to that of the hypernova as an explanation for GRBs. However, two lines of evidence have recently pointed towards the correctness of the hypernova theory. The first is that, as we have seen, very massive stars live for only a few million years, and so the hypernova will occur while the star is still embedded within the gas cloud from which it formed. It has recently been found that many GRBs do indeed occur within star-forming regions. The second line of evidence is that the hypernova explosion should

Figure 2. Stan Woosley (left) and Andrew MacFadyen who conceived and developed the hypernova model for GRBs. (Image reproduced courtesy of R.E. Rutledge.)

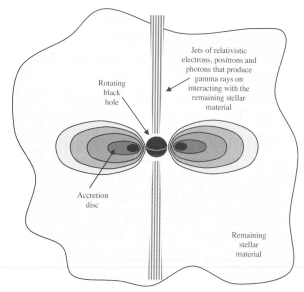

Figure 3. The centre of a hypernova.

sometimes eject sufficient matter and energy to look like a supernova a few days or weeks after the collapse; exactly such a supernova was observed in 1998 in the same direction as GRB 980425, three days after that burst occurred. Similarly, in March 2003, the spectrum of the afterglow of GRB 030329 was identical with that of a powerful supernova. Thus it seems likely that when a gamma-ray burst occurs in the sky, we are witnessing the death throes of a supermassive star as it collapses to form a black hole.

NOTE

1 The term 'collapsar' may sometimes be encountered as a synonym for hypernova. But the term is also used for the collapsing phase of a hypernova, and for objects such as white dwarfs, neutron stars and black holes, and so is avoided here. The term 'hypernova' was coined by Bohdan Paczynski of Princeton University in 1997.

Gamma-ray Bursters
Gamma rays are a form of electromagnetic radiation, just like light and radio waves, but at the short wavelength end of the spectrum. The division between X-rays and gamma rays is usually taken at a wavelength of 0.01 nm (10^{-11} m), with gamma rays having wavelengths equal to or shorter than this limit. This corresponds to frequencies of 3×10^{19} Hz or higher, and energies of 100 keV or more (the electron volt is the energy gained by an electron when it is accelerated by an electric field of one volt, and it has a value of 1.6022×10^{-19} J).

The first gamma-ray burster (GRB) was detected in July 1969 by a pair of the Vela series of spacecraft that were launched to look for clandestine atomic bomb tests. Many more have since been observed and they appear in the sky about once a day. The Burst and Transient Source Experiment (BATSE) on board the Compton Gamma Ray Observatory, and the BeppoSAX and HETE 2 spacecraft have observed many thousands of bursts. Their afterglows are also now regularly detected at X-ray, optical, infrared and radio wavelengths.

The main characteristics of gamma-ray bursts are very variable. The intensity can vary from being by far the brightest gamma ray source in the sky to events at the limits of detection. The bursts can last from a few milliseconds to a minute or more, and the emission can occur as a simple smooth single peak, or as numerous rapidly varying jagged multi-peaks. The energies of the gamma rays range from around 10 keV (wavelength 0.1 nm) to 1 MeV (0.001 nm), with a peak emission around 200 keV (0.005 nm). The distribution of bursts is uniform over the whole sky. A plot of the number of bursts against their intensity, however, shows a tail-off at low intensities compared with that which would be expected if the burst sources were completely uniformly distributed throughout space. Furthermore, the number/intensity curves appear to differ between the short bursts and longer ones, suggesting that there might be two types of objects producing the bursts.

Advancing Astronomical Imaging via the Internet

STEVE WAINWRIGHT

BEGINNINGS

It all began in the summer of 1998. Dave Parkin, a colleague and fellow member of the Swansea Astronomical Society, purchased a Connectix Color QuickCam videoconferencing webcam and told me that he intended to use it for astronomical imaging.

He had seen that there was a relatively small number of people with websites or who contributed to the discussion of astronomical newsgroups on the Internet, who had taken astronomical images with varying degrees of success using such devices. The idea that this could be done was exciting, so I purchased the black and white version of the same Connectix QuickCam. The game was on.

What was particularly exciting was that it was evident from investigating the software that came with the QuickCams that they should be able to take extremely long exposures (Figure 1). We imagined that deep sky objects would come within the grasp of our modest scopes and our webcams. What we didn't know at this time was that the long exposure functions of the software simply didn't work. There was really no need for a videoconferencing camera to be able to take long exposures, so Connectix never fixed the problems that prevented it. However, Dave Parkin was able to quickly show that the QuickCam and his 8.5-inch Newtonian were able to capture quite good images of bright objects such as the Moon (Figure 2).

In the summer of 1998, two articles were published in *Sky & Telescope* that were pivotal to what would happen in the autumn of that year. In June 1998, John Buchanan, a Professor of Geology at Eastern Washington University, published an article in *Sky & Telescope* entitled 'QuickCam Astronomy'. He showed how the circuit board of the Connectix QuickCam webcam could be taken out of its housing and

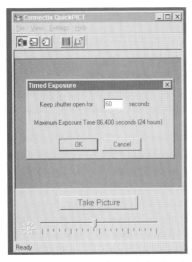

Figure 1. The Connectix Color QuickCam software indicating that exposures of up to twenty-four hours' duration were possible.

Figure 2. Image of the Tycho region of the Moon acquired using a Connectix Color 2 QuickCam. (Image by kind permission of Dave Parkin.)

remounted so that the camera's CCD (Charge Coupled Device) chip could be placed at the prime focus of a telescope. He stated that the QuickCam was not suitable for deep sky imaging and that exposures were limited to no more than a few seconds. He ascribed this to the fact that the camera was not cooled, as were dedicated astronomical CCD cameras. Buchanan was clearly unaware that the long exposure functions of the camera did not work, and he was also unaware, as were many people at that time, that Peltier cooling of an astronomical CCD camera is not essential for the taking of extremely good long-exposure images. It is possible to deal with thermal noise in a way other than reducing it through cooling, by making corrections in software. He described the method he used to obtain his images of the Moon or planets, taking usually a dozen or so images, '. . . hoping that several will be during those split-second moments of good astronomical seeing'. We shall see how things were to change dramatically in the years that followed. Buchanan also made another interesting observation, which was that for about US$200, it was possible to purchase the colour version of the QuickCam. Subsequently, similar amounts of money would purchase high-quality entry-level CCD imagers that would have been considered quite impossible in the summer of 1998.

Two months later, in the August 1998 issue of *Sky & Telescope*, Ron Dantowitz, an astronomy educator at the Boston Museum of Science, published an article entitled 'Sharper Images through Video'. The article concentrated on planetary imaging and the use of video semi-frames that are taken at 1/60 second intervals in the NTSC video standard and 1/50 second in the PAL standard. His idea was to use the best semi-frames captured from Betamax videotape and to combine them into sharp images. He demonstrated the validity of his methods by successfully tracking satellites and imaging them even in broad daylight. For example, in September 1996, he was able to photograph the historic STS-79 mission in which the Space Shuttle docked with the Russian Mir space station. His success in this area attracted the attention of (and alarmed) the 'Men in Black' at the National Reconnaissance Office who paid him a visit. When he was able to show them that by using data readily available to the public he was able to track and image a classified satellite of their choosing, they realized that what they had hitherto believed was only within the capability of the superpowers was actually achievable by a single individual using off-the-shelf components. This was an indicator of the capabilities that amateur astronomers would have quite soon.

What caught my eye was not any of the above, interesting and exciting as it was. I was drawn to two images published in the same article. Here Ron Dantowitz had used a video camera on a 16-inch Schmidt–Cassegrain telescope at f/3.3. The first image was a single semi-frame of the Trapezium area of the Orion Nebula. This image was incredibly noisy and barely recognizable. The second image was an average composite of 200 such semi-frames and showed a relatively noise-free image in which the Trapezium and some of its surrounding nebulosity and stars were clearly visible. This should not be a surprise, as it is well known that the signal-to-noise ratio increases as the square root of the number of images summed or averaged. Thus, for example, if a hundred frames are summed, or averaged, the signal/noise ratio is ten times that in a single frame. Ron Dantowitz had demonstrated to me that deep sky objects were potentially within the grasp of video cameras, if only one had the software to do the job of aligning and summing huge numbers of frames.

To complete the task of convincing me that the related technologies of video surveillance cameras and webcams were much more capable than anyone had recognized, Dave Parkin, using the very limited and faulty software supplied with the Connectix QuickCam Color 2 webcam, produced an image of M13 with his 8.75-inch Newtonian telescope. It was a very poor image by today's standards, but it was clearly recognizable as an image of a globular cluster.

The problem was that I knew what I wanted to achieve. I wanted to take high-quality planetary and lunar images; even more than this, I wanted to take images of deep sky objects, but I didn't know how to do it. In particular, I didn't know how to take long exposures of deep sky objects. Of course the reason for this was that it was impossible. Although the QuickCam software clearly indicated that it should be able to take incredibly long exposures of up to twenty-four hours (Figure 1), the software simply didn't work. However, I had one thing clear in my mind: the Connectix engineers had made a camera, in fact two cameras, the B/W QuickCam and the Color QuickCam 2, which they knew were capable of taking long exposures. What they hadn't done was to write software that could actually exploit this capability. Moreover, because a webcam videoconferencing camera simply doesn't need to be able to take long exposures, there was no point in them fixing a fault in a feature that nobody was ever going to use. I needed help.

I had no significant experience in writing software to control hardware such as an electronic camera. So I was not going to be able to fix the problem for myself, at least not without embarking on a huge learning curve. I knew that the expertise was out there in the world, and that the Internet was the medium by which it might be possible to bring together people with the appropriate knowledge and motivation to solve these problems. Consequently, I decided to establish an Internet forum that would bring together people interested in using unconventional electronic imaging devices such as the QuickCam and surveillance cameras, so that we could share information and ideas. Together, hopefully, we could achieve the 'Holy Grail' of long exposures, so that not only lunar and planetary images would be available to the relatively cheap technologies of the webcam and the surveillance camera, but also elusive deep sky objects would surrender to our low-cost efforts. The only problem was that it seemed impossible.

I searched the Internet and found a handful of people who were using webcams or surveillance cameras to make astronomical images. There was a very active Francophone mailing list called Astrocam, but I needed a group who would communicate in English. I explained my intention of setting up an Internet forum so that we could exchange ideas and results, and I invited the people I found to become members of the forum. On October 15, 1998, QCUIAG was born. QCUIAG is an acronym for QuickCam and Unconventional Imaging Astronomy Group. I ran this group as a manual mailing list and web archive of messages. During the first ten months, the membership grew to about forty and most of the effort went into planetary and lunar imaging. While everyone wished that deep sky imaging could have been achieved by making long exposures, everyone lamented the fact that this function of the software didn't work and so long exposures were impossible.

A BREAKTHROUGH

Fortunately, however, an American member, Dave Allmon, didn't know that long exposures were impossible! Dave made a remarkable discovery. He noticed that sometimes (and quite frequently) when using his B/W QuickCam, the Connectix software 'broke' spontaneously. At least, that is how he described the phenomenon he had observed. When his software 'broke', he found that the bulb shutter function of the camera software now actually worked. It was possible to press a button to open the shutter and to press it again later to close the shutter. Long exposures were within our grasp. However, they depended on the Connectix software 'breaking' spontaneously. I was excited and said that it was obvious that what was needed was a way to 'break' the software at will, so that at least the bulb shutter control function could be used to take deep sky images. I pressed Dave to find a way to 'break' the Connectix software.

Dave Allmon is an expert programmer and a member of the Linux fraternity. Linux is an alternative operating system to Windows and is characterized, among other things, by a community of users who develop software for this platform and make it available to other users. As it turned out, Dave discovered that software had been written under Linux to control the B/W Connectix QuickCam. As this software was written to run under Linux, it did not make use of the Connectix software, or the Video for Windows drivers. This Linux software was capable of 'talking' directly to the camera and of allowing it to take true long exposures. Dave duly set about writing a pair of Windows programs called QCV2 and Darkgen. QCV2 used the Linux techniques and controlled the QuickCam, allowing it, for the first time, to take true long exposures under Windows. Darkgen allowed the averaging of a number of such images to increase the signal-to-noise ratio in the final resulting image. Remarkably, the only hardware modification required was for a single wire to be cut to disable the anti-blooming function of the camera and allow true long exposures to be taken.

In the late summer of 1999, Dave Allmon took his scope, a computer and his B/W QuickCam on a trip to the mountains. When he returned, he posted a series of images of M31, the Andromeda Galaxy, to QCUIAG (Figure 3), demonstrating that the B/W QuickCam was capable of taking long exposures and imaging deep sky objects. Dave had shown that the deep sky is accessible to a lowly webcam, and with the very camera that just over a year previously John Buchanan had said was not suitable for taking long exposures. I started to think that nothing is impossible.

Membership of QCUIAG continued to grow and when the group was just over a year old I put the mailing list under the control of a

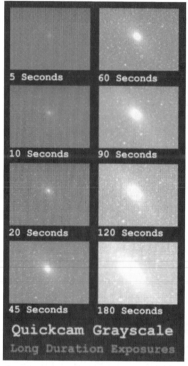

Figure 3. A series of long exposures of M31 taken by a B/W Connectix QuickCam. (Image by kind permission of Dave Allmon.)

Listbot. The automation of the mailing list liberated me from the manual maintenance of the list to spend more time developing the QCUIAG website (www.qcuiag.co.uk) and taking images of my own.

In the meantime, Dave Allmon and others were using his QCV2 and Darkgen software, and they were taking truly remarkable images of deep sky objects with this little B/W webcam (Figures 4–7), while others were taking better and better images of the Moon and planets using the increasing number of videoconferencing cameras that were becoming available. Other QCUIAG members were achieving better and better results with surveillance video cameras. Progress was good and satisfying.

In 1999–2000, Colin Bownes produced Vega, a brilliant piece of software designed to capture BMPs, FITS images and AVIs from

Figure 4. M31, the Andromeda Galaxy, and its companion galaxy M110, by Dave Allmon, using a B/W Connectix QuickCam.

Figure 5. The Horsehead Nebula in Orion, by Dave Allmon, using a B/W Connectix QuickCam.

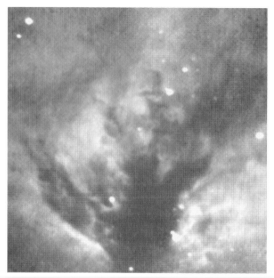

Figure 6. The Flame Nebula in Orion, by Dave Allmon, using a B/W Connectix QuickCam.

Figure 7. The Crab Nebula (M1), by Dave Allmon, using a B/W Connectix QuickCam.

webcams and surveillance cameras. This software evolved over time to meet the challenges that were to follow.

SOFTWARE TO SYNTHESISE LONG EXPOSURES BY ADDING TOGETHER SHORT EXPOSURES

In the spring of 2000, Juergen Liesmann published deep sky images taken with a low-light surveillance camera, something that would have been considered absolutely impossible only a few months before. What was even more remarkable was that the deep sky images were taken with a video camera capable of taking exposure no longer than 1/50 second. Juergen's software, written in Java and running under Linux, was able to sum large numbers of video frames, each of 1/50 second exposure, into deep FITS (Flexible Image Transport System – the image format used by professional astronomers) files. Off-chip video integration was well and truly under way. Juergen ported his program over to Windows, but the Java environment in which the code ran was too much for some of the humbler computers in use by the group. However, Juergen had demonstrated that long exposures could be synthesized from the summation of very short exposures, so here was another route into deep sky imaging with video. 'Long' exposures were no longer in the exclusive realm of astronomical CCD cameras or the humble B/W QuickCam. By summing image files, Juergen was doing what the professionals had been doing for some time, but on a huge scale (Figure 8).

In the summer of 2000, I visited the COAA (Centre for Observational Astronomy in the Algarve) observatory in Portugal. I convinced Bev Ewen-Smith of COAA that it would be worth writing a Windows-based program from scratch, to do off-chip video integration and synthesize long exposures from video cameras in order to make images of deep sky objects. It was thus that the AstroVideo software from COAA was born. A very productive collaboration was established between Bev and myself and AstroVideo has evolved at a phenomenal rate to support all of the subsequent developments made by QCUIAG. What follows here in terms of software development is largely the story of the development of AstroVideo. However, it is important to know that within QCUIAG, various other members were simultaneously developing software for image capture. Programs such as Vega by Colin

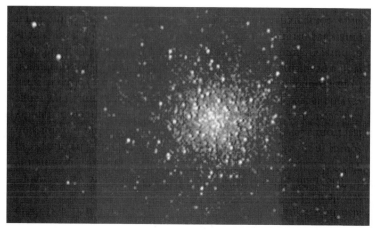

Figure 8. The Globular Cluster in Hercules (M13). 10,000 frames from a 0.00015 lux video surveillance camera captured, aligned and summed by AstroVideo.

Bownes and K3CCDTools by Peter Katreniak are just two examples of the brilliant software innovations to be produced.

My image of the Horsehead Nebula and NGC 2023 (Figure 9) was taken by AstroVideo, which automatically aligned and summed video frames, produced by a 'second-generation' surveillance camera. This camera, from the Mintron stable, is itself capable of integrating the equivalent of 128 frames or 2.5 seconds, exposure into internal camera memory while continually outputting a PAL signal that is updated every 2.5 seconds. The discovery of this kind of camera was a significant advance for QCUIAG and has subsequently been badge-engineered by a number of astronomical equipment suppliers.

The fact that one can synthesize long exposures by adding together thousands of short exposures into 32-bit deep FITS files is remarkable. Very faint parts of the object being imaged gradually build up into significant values within the accumulating FITS file and may then be made visible at the same time as the brighter parts of the image by various non-linear post-capture image-processing algorithms. Moreover, as the final image comprises the sum of thousands of individual video frames, it is almost noise-free because, as previously mentioned, the signal-to-noise ratio increases as the square root of the number of frames that have been summed.

On the face of it, another, even more remarkable, fact is that the cameras used are uncooled. Classical astronomical CCD cameras are Peltier-cooled, to eliminate or reduce the problems associated with thermal noise in a long exposure. The fact is, however, that as long as thermal noise, accumulating randomly in the CCD during an exposure, does not cause any part of the image to become saturated, it can be removed by the subtraction of a dark frame of the same duration. (A dark frame is an image taken with the front of the telescope or lens covered to prevent the entry of any light. The resulting 'dark frame' only contains noise, largely of thermal origin.) When a long exposure has been synthesized from the summation of huge numbers of video frames, each a maximum of 1/50 second, thermal noise is not a problem, for there has been insufficient time for significant amounts of thermal noise to accumulate during the exposure of each frame. However, the small amounts of thermal noise to be accumulated in the CCD pixel wells actually assists in the quantization of very faint parts of the object being imaged, which may not yield sufficient photons to allow quantization to occur. However, the combination of thermal noise plus

Figure 9. The Horsehead Nebula. A long exposure synthesized by AstroVideo, using a Mintron second-generation surveillance camera accumulating exposures equivalent to 2.5 seconds into camera memory while outputting a continuous composite video signal updated every 2.5 seconds.

photoelectrons may be sufficient to record the arrival of the photons. The end result, of the off-chip video-integration process achieved by capture software such as AstroVideo, is a relatively noise-free image with a surprising dynamic range.

The discovery of second-generation video cameras with greatly enhanced sensitivities, such as those in the Mintron range, gave even more power to the process of off-chip integration and the synthesis of long exposures. A paradigm shift was to take place that would allow software such as AstroVideo to exploit the increased sensitivity inherent in medium-length exposures, together with the fact that the exposures are still relatively short, and so avoid star-trailing due to tracking problems.

DEALING WITH SMALL AMOUNTS OF IMAGE DRIFT

Imagine a situation where the telescope tracks badly because it is not perfectly polar-aligned and, moreover, also has significant periodic error in the drives. If a single long exposure is taken, the result will show star-trails due to the poor tracking and possibly other wanderings due to the periodic error. The classic solution to this has either been to guide manually using a guide-scope or to use autoguiding with a CCD guide-camera via a computer, sending guiding signals to the drives.

The AstroVideo software allows for very slight shifts in the positions of objects such as stars in each new frame as it is captured. By a process of cross-correlation in a specified area of the frame, slight changes in the position of objects are detected and the incoming frame is offset by exactly the right amount so that when it is summed into the 32-bit accumulator FITS file, the stars sit down exactly on top of the same stars in the accumulator FITS file. The result is a final, synthesized long exposure with pinpoint stars, whereas a single true long exposure would have trailed star images.

Alternatively, AstroVideo can use the same cross-correlation procedure, on a specified area of the image, to detect slight movement of objects in each incoming frame, and then send guiding commands to the drives, correcting for the image movement.

DRIFT INTEGRATION: ELIMINATING THE EFFECTS OF LARGE-SCALE DRIFT WITH CAMERAS ON UNDRIVEN MOUNTS

Imagine pointing a video surveillance camera at the sky on an undriven tripod (Figure 10). The resulting image shows star trails (Figure 11). However, with Drift Integration enabled, pinpoint stars are obtained (Figure 12). This is how it works. To begin with, two calibration images are taken by integrating a number of frames into each image. The second of the two images is captured when the starfield has drifted a significant amount across the field of view. Then these two FITS images are used by AstroVideo to calculate the rate and direction of drift when a reference star has been selected. In the FITS file header the exact time of capture of the image is recorded, so the software can use this information to calculate the rate of drift. The software is then set to capture a series of integrated images as the object of interest drifts across the field of view. As each new image is captured, AstroVideo calculates the amount of positional offset that needs to be applied to this frame so that the stars in this frame will sit down exactly on top of the same stars in the accumulator FITS file. The same method can be

Figure 10. A video surveillance camera mounted on an undriven tripod.

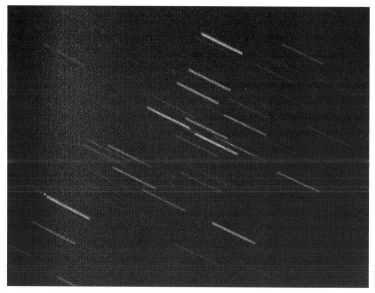

Figure 11. Image of Coma Berenices synthesized by summing 2000 video frames.

Figure 12. The stars of Coma Berenices. 2000 video frames summed by AstroVideo with drift integration enabled.

used to capture a long strip across the Moon as the Moon drifts across the field of view. All of this is done by dead reckoning, as opposed to the cross-correlation techniques described earlier, to compensate for slight drift in the position of the object being imaged by the summation of medium-length exposures by off-chip video integration.

CAPTURING THAT MOMENT OF PERFECT SEEING

You will recall that John Buchanan had said in his *Sky & Telescope* article that his lunar imaging technique involved taking usually a dozen or so images, '. . . hoping that several will be during those split-second moments of good astronomical seeing'. AstroVideo was further developed to increase, by orders of magnitude, the probability of capturing that moment of perfect seeing when taking images of the Moon. Using a webcam or a surveillance camera attached to the telescope, an AVI movie file is captured. At a resolution of 640 × 480, about 900 frames are captured in a minute. I routinely acquire AVIs of up to three minutes' duration of a particular feature on the Moon. AstroVideo then analyses each frame in turn for high spatial frequencies (an indicator of sharp images). The software can then present the user with the sharpest (or the least sharp) image (Figures 13 and 14).

Figure 13. The least sharp frame from a 60-second AVI of 900 frames, captured and analysed by AstroVideo. (Image by kind permission of Derek Francis.)

Figure 14. The sharpest image from a 60-second AVI of 900 frames, captured and analysed by AstroVideo. (Image by kind permission of Derek Francis.)

During the period of AVI capture, there is a much better chance of capturing that fleeting moment of good seeing than John Buchanan had when he acquired a dozen or so images, hoping that he had captured the magic moment of perfect astronomical seeing. During the recording of a 60-second AVI movie file of 900 frames, the frames are saved at a rate of fifteen frames per second. However, the electronic shutter speed could be as fast as 1/1000 second or less depending on the telescope and camera being used. There is a very high probability of capturing a moment of very good seeing under these conditions. The recording of AVIs of longer duration will have commensurately higher probabilities of capturing a very sharp image. AstroVideo can sort through 900 images in seconds, analysing them for high spatial frequencies and sorting them accordingly, a task that would in practice be impossible to do by hand and using visual judgement.

MODIFYING THE CIRCUITRY OF CHEAP WEBCAMS AND SURVEILLANCE CAMERAS TO PRODUCE CAMERAS WITH TRUE LONG-EXPOSURE CAPABILITIES

In the late summer of 2001, a then new member of QCUIAG, Steve Chambers, made another breakthrough. Steve had studied the circuitry of a CMOS- (Complementary Metal Oxide Semiconductor) based webcam, and had worked out how to make a modification that allowed the camera to take true long exposures. The problem with the CMOS device was that it was incredibly insensitive. Moreover, due to the way that charge leaks away quickly from the pixel wells of a CMOS sensor during an exposure and due to photon capture, the longest useful exposures that could be obtained from this device were of about 5 seconds' duration before the image started to saturate. Nevertheless, using AstroVideo to capture and sum 60 groups of 5 × 5-second exposures, an acceptable image of M13 was produced from this device. However, within a month, this CMOS-based camera had been relegated to the status of a curiosity when Steve Chambers successfully modified the circuit board of a Philips Vesta Pro CCD-based webcam. This camera was capable of taking exposures of any length, and was quickly improved to have the CCD amplifier turned off during the exposure and turned back on just before download of the image. This eliminated a phenomenon well known in astronomical CCD cameras where electroluminescence from the amplifier produces an interfering glow on one side of the image. At about the same time, Jon Grove (another member of QCUIAG) produced an analogous modification of a surveillance camera circuit board. The details of the modifications can be found by following links from the QCUIAG website. The circuit modifications required to convert webcams or surveillance cameras to true long-exposure astronomical imagers are not trivial. They require modifications to a circuit board populated by tiny surface-mount components. A dissecting microscope provides the best environment in which the minute soldering modifications can be made. These modifications involve the cutting of component legs and circuit tracks and the soldering of wires to connect to extra support circuitry that has to be built (Figure 15).

It was very gratifying to find that within just over a year of the

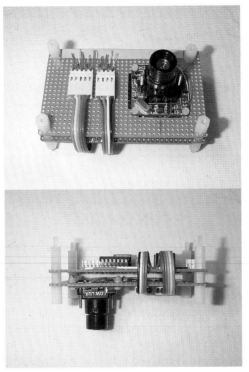

Figure 15. Modified surveillance camera. The extra circuitry is mounted on additional boards before being mounted in a sturdy project box. Layout design by Derek Francis.

invention of these camera modifications, several cameras of related design had been successfully modified, and SAC Imaging in the USA and Perseu in Europe were manufacturing and marketing modified cameras by suitable agreement with the inventors of the modifications. This has very significantly reduced the cost of entry-level astronomical CCD imaging, and makes it possible for people without the necessary micro-soldering skills to own a modified webcam.

The relatively low cost of modified webcams as astronomical imagers is due entirely to the economies of scale involved in the manufacture of webcams that are sold worldwide in their millions, as compared with classical astronomical CCD cameras that have a market measured in hundreds, or at the most thousands, worldwide. The cost of the additional components required is only a few pounds, and additional, modest labour costs are incurred for the commercially modified cameras.

The modified webcams produce one-shot colour imagers that are capable of yielding very satisfying results (Figures 16 and 17). It is now possible for amateurs to invest relatively small sums in either the modification or purchase of modified webcams or surveillance cameras before they invest much more significant sums on the purchase of dedicated astronomical CCD imaging cameras. It is likely that more of the dedicated astronomical cameras will be purchased once astronomers have obtained satisfying results from entry-level devices. This has to be good for the industry as a whole.

At the time of writing, new cameras are being developed by these commercial companies in collaboration with the QCUIAG inventors of the long-exposure modifications. Concept cameras such as widefield imagers and self-focusing cameras are being developed by other QCUIAG members, making use of existing technologies for precise movement and location, such as those found in floppy-disk drives.

The Internet has made all of these advances possible. I have been able to use the Internet to bring together individuals with the appropriate knowledge and skills and a common interest in facilitating the development of low-cost astronomical imaging hardware and software. At the

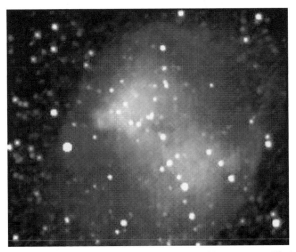

Figure 16. Image of the Dumb-bell Nebula (M27) produced by a SAC7 modified webcam. A number of medium-length exposures were captured and summed by AstroVideo.

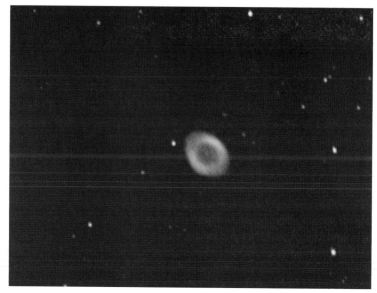

Figure 17. Image of the Ring Nebula (M57) produced by a Perseu modified webcam. A number of medium-length exposures were captured and summed by AstroVideo.

time of writing, there are some 4,300 members of QCUIAG worldwide and the membership is growing. The achievements of the group have been remarkable in the space of five short years, and the QCUIAG website at www.qcuiag.co.uk shows many aspects of the group's work. The developments over the next four years can only be the subject of speculation. However, one thing that can be guaranteed is that they will be exciting. What will be the next significant advance made by QCUIAG? We just don't know!

ACKNOWLEDGEMENTS

I should like to thank Dr Jo McSweeney for reading and criticizing the manuscript and for making numerous helpful comments and suggestions.

The First Planetarium

MICHAEL T. WRIGHT

Mechanical models illustrating astronomical phenomena have a very long history. With all the resources of modern manufacturing, and still more in our age of computer simulation, it is easy to take such devices for granted: 'There are people who can do these things.' To those of earlier generations, who relied wholly on the use of simple mechanical arrangements and the work of their hands, the design and construction of a model showing, for example, the motions of the Earth and Moon about the Sun, or of all the planets in the Solar System, called for rare ingenuity and skill.

EARLY CLOCKS AND ASTRONOMICAL MODELS

The 'orrery', a more or less elaborate model of the Copernican system, used to be a fashionable teaching aid. The name was coined for an early, but not the first, example. It was made by John Rowley for the Earl of Cork and of Orrery in 1712, but the tradition is far older than this indicates. These models almost always depend on the use of toothed wheels, and of course this 'clockwork' grew out of the clock-making tradition. Mechanical clocks appeared all over Europe from about 1300, and the designers of the monumental dials of many of the earliest examples seem often to have been more concerned to present some sort of mechanized model of the cosmos than a convenient means of telling the time.

Since the motions to be modelled are complex, and their periods are awkward, any adequate representation has to be complicated. In many cases this early astronomical dial-work was remarkably sophisticated even though the clock driving it was rather primitive and its timekeeping was not precise. The famous clock of Richard of Wallingford (died 1326) is a good example. A fine reconstruction has now been erected where the original once stood in St Albans Abbey.

Elements of the tradition go still further back. Medieval Arabic culture had water clocks with elaborate displays, of which at least some details had been inherited from earlier Hellenistic culture. The Romans had adopted that Hellenistic culture wholesale, and fragments of a few of their 'anaphoric' clock dials survive. In these, a star map was drawn on a disc which rotated to show the diurnal motion. A ring of holes was drilled through the disc along the circle representing the ecliptic, and a marker was moved from hole to hole, perhaps every few days, to show the current place of the Sun. The dial was turned by a water clock, which was in principle very simple. It depended on the rise of a float in a vessel being filled by a steady trickle of water. The vertical motion was transmitted to the dial by a cord wrapped on a drum.

There are surviving books that tell us about these and other automata from Hellenistic culture. Quite a range of effects was achieved by very simple means such as those just mentioned. Motion was derived from the fall of weights or the flow of water, and regulated by letting water, air or a granular material (such as sand) pass though a narrow opening. It was usually transmitted, if need be, using cords wrapped on drums. Very rarely, toothed gears are suggested.

So, although some ancient writers mentioned planetaria or similar astronomical models, scholars have tended to think that these too were probably quite primitive devices, if not wholly fantastic, like the mythical self-moving tripods of Hephaestus. The famous Roman author Cicero, writing in the first century BC, mentions two early examples brought from Syracuse when the Romans took the city in 212 BC, and associated with the celebrated mathematician Archimedes who died at the end of that siege. In another place he mentions an instrument made by Posidonius of Rhodes, with whom he himself had studied. Cicero offers nothing in the way of technical detail and, until recently, it has been possible for commentators to explain the description of elaborate displays, which imply complicated mechanism, as literary conceits. It took time for the solid evidence of archaeology, which showed that ancient mechanics had achieved far more than modern writers had imagined, to be accepted.

AN INTRIGUING DISCOVERY

In 1900, sponge divers discovered an ancient shipwreck just off the small Greek island of Antikythera. The most prominent items in the cargo, which first attracted their attention, were bronze and marble statues. They reported their find, and the material recovered from that wreck was taken to the National Archaeological Museum in Athens. With good reason it became known as the Antikythera Treasure, and it included some of the finest objects now displayed in the Museum. The wreck had lain under the sea for two thousand years, and it is not surprising that many of the statues were recovered incomplete or in pieces. Some of these must have been mere lumps, perhaps just showing by their colour that they might contain bronze. It was in the Museum, as the Treasure was sorted out, that something remarkable was noticed about one particular lump: small gear wheels and lettering showed on its surface. This little artefact became known as the Antikythera Mechanism.

When it was first discovered, it was hard to make out what this instrument might have been. There was a suggestion that, because it had come from a shipwreck, it might have been a navigational instrument, but as its detail became clearer, this seemed less and less plausible. It is far too complicated to fit into our understanding of the simple navigational techniques of ancient seafarers.

With the realization of the Mechanism's internal complexity came another problem: people were not prepared to accept that such a thing could have been thought of and made in the ancient world. For generations, children had been taught at school that the Ancient Greeks were interested in pure thought, philosophy and mathematics, and that practicalities were despised. This picture is easily shown to be incomplete; it ignores, for example, the intense interest in mechanics of men such as Archimedes. Nevertheless, it was a strongly held belief. There was talk about the Mechanism being a later clock accidentally dropped on to the ancient wreck, despite the facts that it was like no other clock that anyone had ever seen, and that the style of the lettering on it showed it to be as old as the wreck.

Incidentally, the well-established techniques for dating lettering, together with the other means of dating items of the cargo, all point to the Antikythera ship having been wrecked during the first century BC.

This allows us to state that the Antikythera Mechanism is the oldest geared mechanism in the world. The fragmentary inscriptions do more; they offer some clues as to what the Mechanism might have been meant to do. There are graduated dial rings, one representing the Zodiac and another marked out with the days and months of the year. There are other inscriptions concerning the events of the calendar and natural phenomena, perhaps the movements of the planets or the pattern of the weather. This most ancient of mechanical devices was, then, in the tradition of astronomical models that we are discussing.

We can now see why the Mechanism should have been found in the context of the rest of the ship's cargo: statues; fine pottery and glass; fine furniture (of which just the bronze fittings survive); and the inevitable collection of amphorae, the all-purpose containers of the ancient world. This was a cargo of luxury goods, probably from Greek-speaking regions of the Eastern Mediterranean and intended to grace the villas of the rich in or around the city that ruled the civilized world – Rome. Had it not been lost, the Mechanism might have become a rare and precious curiosity in the home of some rich Roman.

FIRST THOUGHTS ON THE ANTIKYTHERA MECHANISM

An understanding of the Mechanism itself was, however, slow to come. Quite probably it suffered accidental damage in the shipwreck, but nearly 2,000 years of lying in seawater was bound to play havoc with such a delicate instrument. What remains is incomplete, crushed and twisted. The thin bronze sheet of which most parts were made has become completely converted to corrosion products, and the surfaces of the wheels and the spaces between them have been filled by stony deposits. Some small fragments of wood, seeming once to have been parts of a box surrounding the mechanism, have shrunk and crumbled on drying out. We are left with a handful of separate pieces of what was evidently once a highly intricate device made with great skill. The question is, do we have enough of it to show what it was and for us to understand what it was made to do?

A great step forward was made in the early 1970s when Dr Charalambos Karakalos, of the Demokritos Laboratory of the Greek Atomic Energy Authority, first subjected the fragments to radiography.

Figure 1. Antikythera Mechanism, fragment A, front, shown approximately 50 per cent of actual size.

Although there is much detail on the surface, this work revealed a great deal more that cannot be seen directly. It is only through radiography that we know that the largest fragment contains the traces of about thirty small toothed wheels, that we can work out how they are arranged, and can make an attempt to count the number of teeth on each. Working with the plates prepared by Dr Karakalos, Professor Derek Price of Yale University showed that the wheel-work embodied a well-known astronomical period relation, and, working from this, was able to offer a reconstruction that accounted for many features of the Mechanism.

Price published his account in 1974. Most of the bronze gear wheels had been mounted on a central bronze frame plate, which was fitted into a box. Either side of this cluster of wheels were two dials which formed the front and back of the box. One dial, which seemed the more impressive one and which Price called the 'front dial', had two big concentric rings: the inner ring, divided into 360°, represented the circle of the Zodiac, with the names of the constellations written in, while the outer one showed the days and months of the year according to the Egyptian calendar, which was in common use at the time.

Price suggested that the front dial was equipped with two indicators, for the mean positions of the Sun and the Moon; since the Sun moves around the Zodiac in one year, the Mean Sun pointer could also indicate the date. The two were interconnected by gearing with the overall ratio 19:254, a good approximation to the mean length of the sidereal month in years which the Greeks had known for centuries and probably learned from Babylonian astronomy. More surprisingly, Price claimed that these motions were combined by the use of a differential gear, and the result, the period of the mean synodic month, was displayed on one of two dials on the opposite face. The function of the other dial on this face was uncertain, because the inscription on the dial itself is unclear and the system of gearing driving its pointer is incomplete. The whole thing was worked by turning a knob at the side.

Figure 2. Antikythera Mechanism, reconstruction by Michael T. Wright, stripped to show its correspondence to the original fragment A (shown in Figure 1).

Figure 3. Antikythera Mechanism, partial reconstruction by Michael T. Wright. The front dial is a planetarium, with pointers for Sun, Moon, Mercury, Venus, Mars, Jupiter, Saturn and the date.

Figure 4. Antikythera Mechanism, partial reconstruction by Michael T. Wright. View showing the case and the knob by which the mechanism was worked.

A FRESH APPROACH

In 1983, I began work on another ancient geared mechanism that my museum acquired. While the Antikythera Mechanism is the oldest geared device in the world, we believe that the Science Museum's instrument is the second oldest. It too has an astronomical connection; one face is designed as a portable sundial, while on the other, driven by the gearing inside, is a display of the phases of the Moon and (according to my reconstruction) the positions of the Sun and the Moon in the Zodiac. It is later and very much simpler than the Antikythera Mechanism, but the two have much in common; they could be thought of as belonging to a single tradition. Therefore my research led me to look much more closely at Price's account, and I became determined to study the Mechanism for myself.

I carried out this research with the late Professor Allan Bromley of the University of Sydney. Together we examined the fragments, measured and photographed them, and made many radiographs. An ordinary radiograph is a shadow picture, and it cannot tell you at what depth within the mass any particular feature lies. The gears are arranged in a number of closely-packed layers, and it was vital to determine the depth of each. For this I made an apparatus for use with the X-ray tube so that we could exploit the technique of linear tomography. Using it, we obtained series of radiographs which are, in effect, cross-sections at different depths.

It was easy to demonstrate that Price's reconstruction of the fragments was wrong in many ways. Most of these concerned minor details, which Price had either missed or mistaken. A few points, however, were serious; there were features that actually contradicted Price's scheme. The worst of these was the observation that, in the very heart of the mechanism, the gearing simply was not connected as Price suggested, which made nonsense of his reconstruction. Then, going through his account point by point, we realized that very little in it was really certain.

Everything, therefore, had to be questioned. Maddeningly, it has proved very difficult to replace the rejected arrangement with a more satisfactory one; it is far easier to demolish than to build! The picture has, however, begun to clear little by little. Provisionally, much of

Price's scheme remains, but there have to be certain modifications and additions.

I saw a way to begin solving this puzzle when I realized that the place at which we have to revise the arrangement of the wheel-work, as mentioned above, is just the point at which the front and back halves of the gearing are interconnected. In other words, it was possible to start from that point and to work outwards in both directions, treating the two halves of the instrument almost as independent problems. If a logical, consistent scheme could be devised for either half it might lead to a solution covering the instrument as a whole, but in any case it was unlikely that it would need radical revision in the light of the reconstruction of the other half.

With this idea in mind, I found that my observations of the 'front' half of the mechanism fell together in a most satisfying way. Price's Sun and Moon motions still made sense, and in fact the revision of the gearing scheme made it easier to obtain these functions than before. But there are details that suggest that the display was more complicated and more interesting than mere indications of the places of the Mean Sun and Mean Moon; there is evidence of epicyclic gearing.

THE IMPORTANCE OF EPICYCLES

In the first century BC, when the Antikythera Mechanism was made, Greek astronomy was still developing towards its culmination in the work of Claudius Ptolemy (second century AD), but the principle of using epicycles had already been established (see box for more detail, p. 214). Applied to the mean motion of the Sun, an epicyclic arrangement provided a way of modelling exactly the solar theory of Hipparchus (second century BC), itself a very good approximation. The lunar theory of Hipparchus, limited in its usefulness but the best that existed before it was elaborated by Ptolemy, could be modelled in the same way. Apollonius of Perga (third century BC) had already applied epicycles to planetary motion to provide models which, although so imperfect that Hipparchus declared that they did not satisfy him, were again the best that were devised prior to the work of Ptolemy.

The Mechanism retains traces of an epicyclic platform that rotated at the rate of the Mean Sun. This could have provided a model of the solar theory of Hipparchus, or a mechanism to show the motion of one or both of the inferior planets, Mercury and Venus, as envisaged by Apollonius. I have devised a scheme, consistent with the evidence, that includes all three possibilities together: on the one platform there is an epicyclic mechanism driving three pointers, for the Sun, Mercury and Venus. A fourth hand, for the date, is fixed to the platform itself and rotates uniformly.

That is about as far as the physical remains can take us, but the designer of such a model would presumably have been interested in making it as complete as possible. I have gone on to show that it is possible to add further mechanism in the same style, to model the lunar theory of Hipparchus and to show the superior planets Mars, Jupiter and Saturn using the same system as for Mercury and Venus. The additions are still consistent with the original fragments, the instrument was not difficult to make using only simple tools, and it works. The additions are contained in extra layers that lie over those modelled on the parts that remain, and they are held in place by a wooden case developed to correspond to the traces of wood found in the original. The whole instrument is constructed in such a way that, if thrown into the sea for 2,000 years, it might well collapse and be reduced to something very like what now survives of the Antikythera Mechanism.

My reconstruction involves extra gear trains. I have chosen these so that each gives an approximation to the required astronomical period (taken, for the sake of argument, from Ptolemy) at least as good as that of the 19:254 ratio found in the original fragments. None of them gives rise to an error in position of as much as one degree in 500 years. (The geometric models adopted for historical reasons do, in fact, give rise to greater errors than this at certain points in their cycles, but these errors die away again at the end of each cycle.) As a demonstration of what is feasible, I believe that I have erred, if at all, on the safe side; I have added forty-one wheels, and since the reconstruction works with this degree of complication, it could certainly work with simpler gearing, giving less precise approximations, if that were thought more probable.

With its display of the whole universe, as understood by ancient astronomers, this planetarium dial is a fascinating thing to watch in motion. We may imagine it being used for teaching, or as an aid to astrology, a 'science' which was growing in popularity at the time and which was certainly taken very seriously by some people. In casting a client's horoscope, the astrologer needed the places of the Sun, the Moon and all the planets at his moment of birth. Using this instrument, he could simply have wound the knob to 'dial in' the date and then (ignoring the approximations of the model) have read off the positions shown on the dial.

In its present state, my model represents a reconstruction of only the 'front' half of the Antikythera Mechanism but, as I pointed out above, the two halves can be treated as practically independent problems. The 'back' half, with its two dials (one above the other on an oblong dial plate), poses a different sort of puzzle. Here the wheelwork, at least for the lower dial, seems more complete, but it is not easy to identify the ratios so as to see what it was intended to do, or to understand how information was to be shown on the dials. Price thought that the pointer on the lower back dial turned once in a synodic month, so that perhaps the dial showed the phases of the Moon. I now think this cannot be right, and I see instead the possibility of adding a display of the synodic month and the Moon's phases to the planetarium dial. (In mechanical terms, this would be only a slight addition.) In any case, whatever the solution for the back may turn out to be, I do not expect it to force me to revise radically my planetarium display for the front dial.

The arrangement of parts of my reconstruction may, however, very well be wrong; I think that, unless a great deal more information comes to light, we can never know in detail just how the Antikythera Mechanism was arranged. The important point is that what remains is consistent with my reconstruction as a planetarium, and by making a working model, using only simple tools, I have shown that such an instrument would not have been beyond the ability of the man who designed and built the Mechanism. We no longer have any reason not to take the reports of planetaria, written by Cicero and later authors, quite literally.

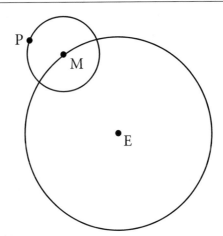

Figure 5. A simple epicyclic system. E is the Earth, at the centre of the larger circle. P is the place of the planet or other body, on the smaller circle – or epicycle – centred at the mean position M. For a planet, P rotates around M in the same direction as that in which M rotates around E. For the Sun, P does not rotate around M, but the direction MP remains fixed as M moves around E. For the Moon, P rotates slowly in the opposite direction to that in which M moves around E.

Greek Astronomy and Epicycles

We think of the Earth and other planets moving around the Sun in elliptical orbits, their velocities varying as described by Kepler's laws. The Ancient Greeks tended to think of the Earth as fixed, with the Sun, Moon and planets moving around it, and sought to explain what they saw in terms of combinations of uniform circular motions.

Apollonius of Perga (third century BC) realized that the apparent behaviour of the planets could be modelled, at least in principle, by an epicyclic system; the planet was supposed to move around a small circle (the epicycle) while the centre of the epicycle moved around a larger circle, the centre of which was the Earth. We can see the combination of these two circular motions as reflecting the combination of the motions of the planet's orbit and the Earth's orbit, both around the Sun.

For the *inferior planets* (those which lie between the Earth and the Sun), Mercury and Venus, the motion around the epicycle reflects the planet's orbit around the Sun, and the motion of the epicycle around the Earth reflects the Earth's orbit. For the others, the *superior planets*, the correspondence is reversed; the motion around the epicycle reflects the Earth's orbit while the motion of the epicycle itself corresponds to that of the planet.

This model is oversimplified because its uniform circular motions are poor approximations to the true motions of the planets. It can reproduce, for example, the remarkable appearance of a planet's retrograde motion, but even if the mean periods are correct the times and positions of the beginning and end of any particular retrograde arc are likely not to be right. For this reason, apparently, the famous astronomer Hipparchus rejected such a model, but no significant improvement seems to have been made until further complications were added to it by Ptolemy. It was, however, the best model of planetary motion available when the Antikythera Mechanism was built, and it was at least easy to build into a working instrument.

Hipparchus (second century BC) developed a solar theory according to which the path of the Sun's apparent motion was a circle eccentric to the Earth, along which the Sun moved with constant velocity. This approximation actually gave results that agreed very well with observation of the Sun's apparent motion, which, as we would recognize, is movement in an ellipse with varying velocity. It was understood that exactly the same effect would be obtained if this large circle was moved to be centred on the Earth, and the Sun was placed on the circumference of a small epicycle which did not rotate. The second arrangement may sound more involved, but it is the easier of the two to translate into a working model.

Compounding uniform circular motions can actually be quite a good way of approximating any periodic motion. In the lunar theory of Hipparchus, the Moon is also placed on an epicycle, but this time the epicycle rotates slowly as its centre moves around the Earth. This model describes the motion of the Moon near syzygy (New Moon and Full Moon) well, but works less well at times in between. Again it is easily mechanized.

FURTHER READING

D.J. de S. Price, 'Gears from the Greeks', *Transactions of the American Philosophical Society*, vol. 64, no. 7, 1974. Reprinted as an independent monograph, New York, Science History Publications, 1975. Also published in Greek, Thessaloniki, Technical Museum of Thessaloniki, 1995.

M.T. Wright and A.G. Bromley, 'Towards a New Reconstruction of the Antikythera Mechanism', *Extraordinary Machines and Structures in Antiquity* (Proceedings of a conference, Ancient Olympia, August 2001), Patras, Peri Technon, forthcoming.

M.T. Wright, 'A Planetarium Display for the Antikythera Mechanism', *Horological Journal*, vol. 144, no. 5, May 2002, pp. 169–173, and vol. 144, no. 6, June 2002, p. 193.

M.T. Wright, 'In the Steps of the Master Mechanic', *Ancient Greece and the Modern World* (Proceedings of a conference, Ancient Olympia, July 2002, forthcoming).

The Ashen Light of Venus

FRED W. TAYLOR

It has long been known that the night side of Venus is not completely dark. Many observers, since Johannes Riccioli of Bologna as long ago as 1643, have reported the emanation of a mysterious glow from the main disc when observing the planet at times when the sunlit side presents itself as a narrow crescent. The glow (shown in Figure 1), known as the 'Ashen Light', is extremely faint, and not always visible. A few observers, most notably Barnard, have never seen it, and doubted its existence, but they are far outnumbered by those who have. They describe it as being dim, dull, red, rusty or brownish, mottled or uneven in its coverage of the night side, and variable in intensity so that sometimes it cannot be detected at all.

There has been much discussion over the centuries of what could be the source of this light (see the summary in Chapter 7 of *Venus*, by

Figure 1. A sketch of Venus by Patrick Moore, using a 15-inch reflector, showing the Ashen Light (its brightness exaggerated for clarity). From *Venus*, by P. Moore, Cassell, 2002.

Patrick Moore, Cassell Illustrated, 2002), but no general agreement. Suggestions have included sunlight scattered quasi-horizontally through the cloud layers, global-scale lightning storms, and some kind of electrically produced aurora analogous to the 'Northern Lights' here on Earth (as well as imaginative suggestions involving celebratory fireworks and what we would now call biomass burning by hypothetical inhabitants of Venus).

Recent research on the atmosphere of Venus has, however, now revealed surprising properties of the surface and the clouds that offer a new explanation, which is almost certainly the correct one at last. We now know that the cloud layers on our sister planet, although extensive, are translucent at visible and near-infrared wavelengths, and it follows that we are seeing through them to the surface, which is glowing faintly with a dull red heat.

The origins of this revelation date from the mid 1980s when astronomers studying Venus detected bright emission from the night side in certain narrow wavelength bands in the near infrared (see the review by the present author in *The Century of Space Science,* ed. J. Geiss *et al.,* pp. 1405–23, Kluwer Academic Publishers, 2002). They show night-side emission similar to the historic reports of Ashen Light in the visible, but much stronger and always present, most prominently at wavelengths of 1.7 and 2.3 microns (see Figure 2). They also show contrast variations which appear to be due to cloud formations, part of the meteorological behaviour of the mysterious lower atmosphere on the Earth's twin. The fact that this had not been seen before was simply due to the fact that no one had looked at these wavelengths. Its discovery, by the late David Allen and his colleagues, occurred when he was using the planet as a calibration target during an observing run to study cool stars at the Anglo-Australian Telescope.

The wavelengths at which the night-side emission can be seen are those which form 'windows' between the strong absorption bands of the gases (principally carbon dioxide and water vapour) which absorb infrared radiation most strongly in the Venusian atmosphere. This implies that the emission must be thermal emission from the lower atmosphere, diffusing through the clouds. Calculations of the emission from a model Venus atmosphere (Kamp, L.W., Taylor, F.W., and Calcutt, S.B., 'Structure of Venus' atmosphere from modelling of night-side infrared spectra', *Nature,* **336,** 360–2, 1988) showed that it was in fact quite possible for infrared radiation to escape through the

Figure 2. Venus through a 2.3-micron filter on a large telescope, showing the illuminated crescent and the infrared version of the Ashen Light on the night-side hemisphere. (The small features at top left are artefacts due to dropouts in the CCD used to record the image.)

clouds. The principal reason is that sulphuric acid droplets have a very high single scattering albedo in the near infrared, so photons are more likely to be scattered than absorbed.

The radiative transfer calculations that explain the emission seen in the 1.7- and 2.3-micron windows also predict that a number of other windows, not observable from the Earth because of absorption in the terrestrial atmosphere, could probe not only the cloud layers on Venus but all the way to the surface. To do this effectively would require a near-infrared imaging spectrometer on a spacecraft making a close approach to Venus. At around this time (1989) the Galileo Jupiter orbiter spacecraft was being prepared for launch, and was to reach Jupiter by means of close fly-bys of Venus and Earth. Galileo carried the perfect instrument for observing Venus, the Near Infrared Mapping Spectrometer (NIMS). Calculations showed that NIMS could achieve a spatial resolution on Venus of as good as 25 kilometres, far better than the Earth-based images. Also, it had a spectral range from 0.7 to 5.2 microns, covering all of the predicted windows.

The results, in February 1991, and shown in Figure 3, were spectacular. The bright regions in the picture correspond to a total cloud thickness about twenty times less than the dark regions. The cloud veil on Venus is not the thick, uniform mass which it had for so long appeared to be. The morphology of the clouds speaks of intense weather activity, featuring large, cumulus-like clumps; exactly what is going on is impossible to say from the stationary snapshot that results from a fly-by, but we can say that the deep atmosphere of Venus must be a very active place.

The window in the visible part of the spectrum, in which the historic observers saw the Ashen Light, was unfortunately just beyond the short end of the wavelength range of NIMS. However, NIMS did observe Venus in the window at 1.05 microns, just beyond the visible in the very near infrared. Figure 4 shows the resulting image, and compares it with a radar map of the surface of Venus, from which it can be seen that the most prominent, i.e. the highest, surface features are indeed visible through the clouds. Since they are viewed by NIMS in thermal

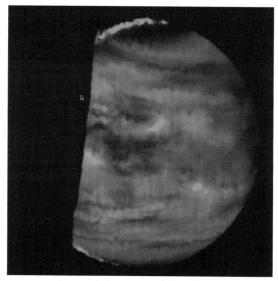

Figure 3. Venus at 2.3 microns wavelength, as observed by the Near Infrared Mapping Spectrometer on the Galileo spacecraft in February 1991. (Image courtesy of NASA/JPL.)

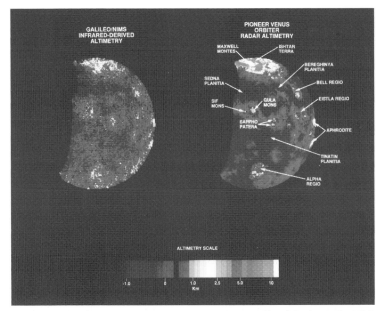

Figure 4. Surface features on Venus seen at 1.05 microns wavelength by the Galileo NIMS spectrometer (left), and by radar from Pioneer Venus (right). (Carlson, R.W., Baines, K.H., Girard, M., Kamp, L.W., Drossart, P., Encrenaz, T. and Taylor, F.W. (1993), 'Galileo/NIMS near-infrared thermal imagery of the surface of Venus', Lunar and Planetary Science Conference XXIV Abstracts, p. 253.)

emission, high features appear dark because they are colder, as a result of the natural fall-off of temperature with height in the atmosphere. A lofty feature like Maxwell Montes, for example, at 11 kilometres above the datum, is more than 100°C colder than the surrounding plains and shows up very clearly in the infrared maps made from space. In order to make the infrared and radar observations directly comparable, both are presented as heights in Figure 4 using the same grey scale. To achieve this, the 1.05-micron brightness has been converted to a height assuming a vertical temperature gradient of 10°C per kilometres, which is the standard adiabatic lapse rate for conditions on Venus and which probably applies, with only small variations, to the lower atmosphere everywhere on the planet.

Of course, it does not automatically follow from the success of a sensitive infrared instrument, in space near Venus, that the human eye could detect the emission from the planet at the shorter, visible wavelengths using an Earth-based telescope. To investigate this, the author's colleague, Dr Guy Peskett, an expert on photonics, has made some preliminary calculations. The biggest uncertainty in these is produced by an apparent lack of detailed, quantitative data about the performance of the eye – in particular, how many photons are required to produce a recognizable response and how this varies across the visible wavelength range. The assumption of a 10 per cent efficiency at a peak wavelength of 0.5 microns, and a Gaussian fall-off to either side of this, leads to approximate equality with the calculated flux from Venus when viewed through a 10.25-inch telescope. M.B.B. Heath, as reported by P. Moore (op. cit.) has made regular observations of the Ashen Light through an instrument with this aperture. If the response of the eye is assumed, more optimistically, to be 10 per cent everywhere across the visible range, the signal-to-noise ratio increases to about 500. On the basis of this, it seems not unreasonable to expect a fully dark-adjusted observer to be able to see the glow from the surface of Venus under good viewing conditions.

In summary, it is possible, and indeed likely, that the famous Ashen Light of Venus is in fact thermal emission from the surface of the planet. The Galileo results and radiative transfer models show that light from the surface of Venus, which is hot enough to glow dull red, can propagate through the clouds. At visible and near-infrared wavelengths, the sulphuric acid aerosols which make up the Venusian clouds are very conservative scatterers, with relatively little absorption when compared with terrestrial water clouds. The many reports of the colour of the night side of Venus as being dull red or rusty lend further support to this idea. The observed patchiness is due to the large-scale structure of clouds in the deep atmosphere. The Galileo spacecraft is currently being fêted for its revelations about its primary objective, Jupiter, as the mission comes to an end with a planned crash into the giant planet. It may inadvertently also have shed light on another of the Solar System's oldest puzzles, during its fly-past of Venus more than twelve years ago.

A Short Guide to Astronomical Art

DAVID A. HARDY

It is fairly easy to find information on how to make sketches through a telescope, especially if you belong to an organization such as the British Astronomical Association (BAA) or the Society for Popular Astronomy (SPA). Certainly this is an excellent way in which to hone your observational powers and drawing techniques, often using a soft pencil and photocopied templates for the oblate shape of Jupiter, the various ring aspects of Saturn, the phases of Venus, Mercury and even Mars (gibbous only, of course, because Mars is further from the Sun than we are), and so on. But perhaps, as I did many years ago, you decide you would like to paint the actual *landscapes* of those worlds. When I did so, around 1950, it was very largely a matter of imagination and guesswork, as telescopes were the only method of observation. Our knowledge and visual impressions of the planets and moons have changed enormously since then; Mars no longer has blue skies and vegetation, Venus has no oceans of soda water, Saturn isn't the only planet with rings – and who could have forecast the volcanoes of Io, the ice-crusted oceans of Europa, the geysers of Triton, the ice cliffs of Miranda? These are the stuff of dreams for artists!

THE FIRST SPACE ARTISTS

Astronomical art, or space art as it is now more usually known (purists insist that a vehicle or figure be included in space art, while astronomical art can be pure landscape), dates back a long way. In 1874 a book was published in England entitled simply *The Moon*, by James Nasmyth and James Carpenter. Nasmyth created accurate plaster models of the Moon's surface, lit them correctly and photographed them against a starry, black background as illustrations. These are probably the first

Figure 1. Eruption on Io (gouache). Something which no space artist anticipated when painting Jupiter's moon before the Voyager fly-bys – active volcanoes. At first these were thought to be relatively cool, caused by sulphur and its compounds, but they are now known to be extremely hot, and generated by internal tidal stresses. An artist needs to be aware of these scientific facts in order to depict such a phenomenon accurately.

examples of true space art. By the early 1900s Scriven Bolton was using a similar technique for the *Illustrated London News*. He and a Frenchman, the Abbé Moreux, worked on a magnificent two-volume book, *Splendour of the Heavens* (1923).

However, their work was eclipsed by that of another Frenchman, Lucien Rudaux, who was born in 1874 and became director of the observatory at Donville, Normandy. He also wrote and illustrated his own books, such as the classic *Sur les autres mondes*. Often he observed the 'limb' or edge of the Moon, where its ravaged surface is seen in profile. So while other artists showed lunar mountains as being steep, jagged peaks, Rudaux painted them as rounded and eroded (not by air or weather, of course, but by eons of impacts by micrometeorites and extremes of temperature). In fact, his paintings, while quite impressionistic, often resemble Apollo photographs. A crater on Mars has been named after him.

Figure 2. Geyser on Triton (acrylics). The artist was at JPL in Pasadena in August 1989 when the images came back from Voyager 2's fly-by of Neptune and Triton. A big puzzle was the dark streaks on the pinkish methane/nitrogen of Triton's surface. They are thought to be caused by geyser-like columns of gas, some of which drift down to the surface causing dark linear deposits. This painting appeared on the covers of *Sky & Telescope* (USA) and *Popular Astronomy* (UK), and also in *The Fires Within* by David A. Hardy and John Murray.

THE OLD MASTER OF SPACE ART

The best-known astronomical painter of all also worked briefly for the *Illustrated London News*, though on architectural renderings, in which he was trained. Chesley Bonestell was born in 1888. The Wright brothers were then seventeen and twenty-one, H.G.Wells was twenty-two and not yet published; yet Bonestell was able to see his visions of men walking on the Moon and probes visiting most of the major planets become reality, for he died in 1986. In 1985 an asteroid, previously unromantically known as (3129) 1979MK2, was renamed after him – a unique honour for a living artist.

Bonestell's first published astronomical art was a series of paint-ings of Saturn from its (then nine) moons for a 1944 issue of *Life* magazine. Arthur C. Clarke was outraged by the comment of a short-sighted editor that 'the figures [of astronauts] are included only to give scale'! Bonestell worked on special effects for films under producer George Pal, including *Destination Moon* and *Conquest of Space*. He went on to lead a team of illustrators, working with scientist-writers under Wernher von Braun and Willy Ley, on a series of articles for *Collier's* magazine, showing how humans could explore space, from a wheel-shaped station in orbit to a fleet of moonships, and later a mission to Mars.

There can be no doubt that they succeeded spectacularly in showing the US public, whose lives had been dominated by Korea and the Cold War, a vision of a new frontier and great glory. In fact, they created a climate in which NASA could begin its work. Von Braun later designed the motors which launched America's first artificial satellite, and the Saturn 5 which took Apollo astronauts to the Moon.

But Bonestell's highly realistic, even photographic oil paintings had another effect. For even though he had a reputation for great accuracy, his moonscapes showed dramatic, towering mountains – so much more inspiring than the flat, drab landscape on which Apollo 11 landed. (Could the disappointment of this reality have been a factor in the public's rapid disenchantment with the space programme?) Through-out the 1950s, science fiction and other illustrators who had never even looked through a telescope produced variations of 'Bonestell's Moon'. Even so, the books which followed the *Life* and *Collier's* articles – *Conquest of Space* (1949), *Across the Space Frontier* (1952) and *Man on*

Figure 3. Mercury and the Sun (digital). Early paintings of Mercury by artists such as Chesley Bonestell showed it as both the hottest and coldest place in the Solar System, since it was believed that one hemisphere was always turned towards the Sun. Mercury's 'year' is actually eighty-eight days and its 'day' 176 Earth days. Mariner 10 imaged only about half of its surface, which appears very Moon-like, though with some features not found on the Moon. NASA's MESSENGER probe is due for launch in 2004, which may result in more paintings of this planet.

the Moon (1953), among others, inspired future generations of space artists, including myself.

THE FOUNDING OF THE IAAA

In 1981 a group of artists from the USA and Canada came together in Hawaii, and formed what is now the International Association of Astronomical Artists (of which I'm currently European Vice-President).

We have around 120 members from many countries, not all of whom are realists; some produce work which is impressionistic, expressionistic, abstract or surreal, but the majority (unlike science fiction and fantasy artists, who work almost purely from imagination) do have a background in astronomy, physics and mathematics, which enables them to interpret accurately the data from observatories and space probes, and convert them into believable scenes. We hold workshops every year or two in some of the most alien parts of planet Earth. My first was in Iceland in 1988, where I was amazed to meet artists who actually 'talked the same language' as myself! Other workshops have been in Hawaii, Utah, Yellowstone Park, St Helens, the Crimea and at observatories and NASA establishments. We hold critiques, exchange notes and tips (now often by e-mail, over our own listserve by which any member can send a message to all other subscribed members) – and have a lot of fun.

Most of the artists would dearly love to visit other worlds; even to take a trip into Earth orbit; one, the Russian cosmonaut Alexei Leonov, who attended the Iceland workshop, has done just that. Another, Alan Bean, has been to the Moon and now sells his paintings to those who can afford them. In fact, space art is a field of rapidly growing interest for art collectors, and the US company Novagraphics (www.novaspace.com) specializes in selling prints and originals by the leading exponents. Space artists paint in oils, acrylics, gouache and markers, and use pens, pastels, coloured pencils or the latest computer technology. But these artists have an advantage over mere technology, for they can travel where machines cannot; and this includes into the past, the future, and faster than light . . .

GETTING STARTED

So how can you go about making a space painting? If at all possible, do try to make sketches of actual rock formations first. (I say *sketches*; if time is short you may have to resort to a 'snap', but there is nothing like spending an hour or two sitting amidst a landscape to absorb its qualities – the textures, the way light bounces off rocks and reflects into shadows, and other subtleties.) If you try to paint rocks and mountains from imagination you may find that they look like lumps of Plasticine!

But, when sketching or looking at references, do bear in mind the types of rocks and geological features that we are likely to find on the other planets of our Solar System (other stars and planetary systems perhaps allow more scope, within scientific probability). This is why we visit volcanic places like Iceland and Hawaii, where there are shield volcanoes like those of Mars, fault valleys, and so on. And it is not enough to paint a landscape with a couple of spheres in a black (or purple or green) sky and call it space art; astronomical art must be informed and as accurate as possible, whatever the style and medium used. You already know at least some astronomy, since you are reading this book; but learn some geology, too, and check your knowledge of the physical characteristics of all the planets, moons and other bodies, including galaxies, nebulae, pulsars, neutron stars, black holes . . .

A common theme in space art is 'Planet A seen from its satellite B'. Nowadays, computer software like *RedShift, Voyager* and especially *Starry Night* make it relatively easy to see how big, say, Jupiter will appear from Europa, and if you are into programming you may be able to produce your own customized program to calculate angular diameters. When I started out, I had to rely on a set of school 'trig' tables. If you don't own a computer you will need this formula: $A = 57.3(D/X)$.

In this, D represents the diameter of the body (planet or moon) in the sky, X is the distance of the observer from the centre of that body, and A is the angular diameter, in degrees, of that body. This works well only as long as the answer is less than 30°. The precise formula involves the trigonometric tangent function: it is $\tan(A/2) = D/2X$. Put another way, the angular diameter of any body is twice the angle whose tangent is $D/2X$.

What do I mean by 'angular diameter'? As you know, we divide the dome of the sky into 180° from horizon to horizon. The Moon's angular diameter is about half a degree in our sky; from the Moon, Earth will occupy 2°, as it is four times bigger. Personally I think of my eyes as a camera, which allows me to view a scene through an imaginary lens, usually with a field of view of about 50°, but it could be longer-focus (say 30°), or wider (say 70°), to give a special effect. But you need to control this; the whole effect of not just the sky but the whole landscape will be changed by this angle. However, taking 50°, then if we are painting on a board or canvas that measures 20 cm × 50 cm, we could use a scale of 1° to 1 cm to give a natural effect. Most people find it surprising that the Moon would occupy only $\frac{1}{100}$ the width of such a picture, but the illusion of the apparently larger size of a rising Full Moon is well known. By an amazing astronomical coincidence, the Moon and the Sun appear almost exactly the same size in our sky – hence total eclipses.

The next thing you have to do is decide where the Sun is. In space subjects, where there is often no atmosphere, light and shade is very harsh, and the shadows are relieved only by the glow reflected from a planet or other object in the sky. So you have to decide at the outset how high the Sun is in the sky, what other bodies will be visible, where they will appear in relation to constellations (if you are using that much detail, which is a good exercise anyway) and, bearing in mind where the plane of the ecliptic will cross the sky area, how much, if any, of the Milky Way will be visible, and so on. You will also need to calculate, based on the height and angle of the Sun, where your shadows will fall and how long they will be.

The easiest way to work out shadows is to draw a vertical pole of the same height as your mountain or whatever, draw an imaginary line from the Sun (or other source of illumination) to the ground, and then draw in the shape of the formation which is casting that shadow. Quite complex shapes can be simplified into cubes or pyramids and treated in this way. The angles, shadows and reflected light of Saturn and its ring system would require an entire lesson in itself!

One dilemma which faces space artists, the answer to which is really down to individual choice, is whether to show stars. No stars appear in, for example, the Apollo photographs taken on the Moon or in space, but this is because of the shortcomings of film, which does not have the latitude to handle a brightly lit landscape and faint objects like stars. If you try to expose for the stars, the rocks become a totally washed-out glare. Even the eye's iris has the same problem, and closes down to a dot in bright light, opening up in dark conditions. Yet the stars *can* be seen if the eyes are shielded, and my view is that here the artist has an advantage, as he/she can show both in the same picture. The same applies to a total solar eclipse seen from the Earth: only an artist can show the faint outer corona in all its glory, as it is seen by the naked eye, without the photosphere around the Moon's rim becoming hopelessly overexposed.

Finally, drawing and painting techniques. Frankly, if you haven't done enough art to know your favoured medium and style, it would be best to take at least a basic painting class at an art school. Generally, you will need to make an accurate drawing on tracing paper first, with all your angles and so on worked out. I often start on a sheet of board which I give two coats of matt black paint. Oils, acrylics, gouache, even pastels – it's all down to personal choice, and I use all of these, sometimes in combination. But for illustration work, nowadays I work mostly on a Power Macintosh G4 1.25GHz MP, with two 120GB hard drives. That again would require a separate article . . . As to a support for your painting, you can use hardboard (smooth or textured side according to choice), watercolour paper, illustration board, canvas, canvas-textured board, again according to personal taste and pocket. A visit to any art shop will enable you to make your choice.

The basic shape of the landscape or object is then painted in white, or a pale colour, and shadows, details and textures built up layer by layer, possibly with the help of an airbrush to add glows, atmosphere or subtle gradations as and when required. Whatever medium you use, it

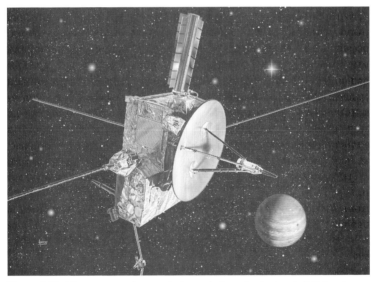

Figure 4. The Ulysses spacecraft passes Jupiter on its looped trajectory. A digital work commissioned by PPARC, this is an example of how an artist can show the probes and satellites which themselves send back images of celestial objects. Obviously it is impossible to obtain photographs of these *in situ*, so only the artist can depict them.

is possible to apply 'glazes' of transparent colour over a lighter base, so that it glows in a way that solid pigment does not.

There is no substitute for experience, and the best way is just to do it. Good luck!

FURTHER INFORMATION

David Hardy's books, slides (including 6 cm × 6 cm transparencies for reproduction), prints of his work and a Screensaver/Wallpaper for Windows and Mac are available from AstroArt, 99 Southam Road, Hall Green, Birmingham B28 0AB, England (tel. 0121 777 1802). Please send SAE for details. Or visit the AstroArt website (which includes 'Teach-in' pages) at www.astroart.org.

Anyone interested in space art (whether an artist or not) may join the IAAA by contacting David Hardy via the above address. The IAAA website is at: www.iaaa.org.

All illustrations accompanying this article are by the author.

Absolutely Nebulous

FRED WATSON

One of the lighter sides of scientific life in New South Wales is a regular event with the alluring title of 'Science in the Pub'. Its format is more or less what you would expect from the name: a couple of scientists debating a hot topic in their field of expertise in a pub – aided and abetted by a compere, with plenty of beer and audience participation. In the half-dozen or so years of its existence, it has proved to be a very successful combination of entertainment and education, in both metropolitan and regional New South Wales.

You might be forgiven for wondering why such an insightful glimpse into Antipodean social life should find its way into the opening paragraph of a *Yearbook of Astronomy* article. The connection is tenuous – perhaps even nebulous – but there really is one. It is that participants in 'Science in the Pub' must, by tradition, set out their opening pitch in the form of an abstract – which has to be in verse. In the opinion of your humble author, abstracts in verse are a Good Thing, and there should be more of them in the scientific press. So, with that in mind, here is a summary of what this article is all about–with apologies to anyone who is offended by the less refined forms of doggerel. And, to anyone else who thinks this particular abstract might be a left-over from a heavy night at 'Science in the Pub' – well spotted.

Forbidden Lines

> The Universe, a largish place,
> Is blessed with lots of empty space
> Where things go on, behind our backs,
> Things too hot for tabloid hacks.
> Here, atoms, free from earthly pressure,
> Cavort in rare and wanton pleasure,
> And, in their frenzied celebration,
> Emit forbidden radiation.

Back on Earth, it took a while
For scientists to spot the guile
With which such nuclei betray
Their games in distant nebulae.
An unknown substance, it was deemed,
Produced the spectrum lines they'd seen.
They christened it 'nebulium'.
(They should have guessed – it rhymes with 'dumb'.)

Then, in nineteen twenty-eight,
Someone came who put them straight.
A clever chap called Ira Bowen,
Told them things they should have – er, knowen.
'Forbidden lines, you see, become
Permitted, when the pressure's gone.
The secret's there in vacuo –
Nebulium's just plain old O'.

'How brilliant!' his peers exclaimed,
'Young Ira's got nebulium tamed!'

Unfortunately, in the street,
Few people heard of Ira's feat,
And, in the folklore of the sky,
Nebulium's still riding high.
Alas, I fear, it's much the same
With all things done in science's name.
Once ideas get recognition,
They regress – to superstition.

THE CELEBRATED PHAENOMENA OF COLOURS

The tale of 'nebulium' is one of the great detective stories of early twentieth-century astrophysics. It has its roots in the pioneering days of stellar spectroscopy – that marvellous technique of splitting starlight into its component rainbow colours to reveal intimate stellar secrets across the abyss of space. In the 1860s, this newly discovered celestial conjuring trick was at the cutting edge of astronomy. Remarkably, it still is.

Of course, astronomers had been looking at the spectrum of one particular star since the seventeenth century. Newton was only twenty-nine when he wrote:

> I procured me a Triangular glass-Prisme, to try therewith the celebrated Phaenomena of Colours. And in order having darkened my chamber, and made a small hole in my window-shuts, to let in a convenient quantity of the Suns light, I placed my Prisme at his entrance, that it might thereby be refracted to the opposite wall . . .

Newton's discovery that white light is composed of individual spectrum colours was only properly understood when, at the turn of the eighteenth century, Thomas Young recognized that the colours correspond to light of differing wavelength. And then, in 1802, William Wollaston noticed dark lines crossing the solar spectrum – although he rather missed the point when he took them simply to be the boundaries between one colour and the next. But during the ensuing sixty years, with the work of Fraunhofer, Kirchhoff, Bunsen and others, the true significance of the dark lines as chemical messengers from the Sun's atmosphere was recognized. They were absorption lines, dark lines whose position in the solar spectrum coincided exactly with the positions of bright emission lines in the spectra of various terrestrial elements when they were excited by a spark or flame in the laboratory. It was a gigantic step forward, and it paved the way for the birth of stellar astrophysics.

Fraunhofer's name is immortalized in the Fraunhofer lines – the name we give to the solar absorption lines today – and he also seems to have been the first person to observe the spectrum of a star other than the Sun (Sirius, in 1814). But the name we usually associate with the early development of stellar spectroscopy is that of William Huggins, who was born in 1824. From the age of thirty, Huggins was able to support his work in astronomy by means of a modest inheritance. In 1862 he was inspired by Kirchhoff's detailed analysis of the solar spectrum. He promptly fitted the 8-inch (20-cm) refracting telescope at his London observatory with a two-prism spectroscope built with the help

Figure 1. Emission lines in the spectrum of a glowing gas. Each vertical line corresponds to light of a specific wavelength, and has its own distinct colour. The more familiar rainbow-like spectrum of white light exhibits a continuous band of colour from violet to red, representing all wavelengths.

of his friend William Miller, a professor of chemistry at King's College. Together, these two men then embarked on a spectroscopic romp through the heavens – observing the Sun, Moon, planets and, most significantly, the stars.

Of course, it was only the brighter stars that yielded their rainbow secrets to the crude equipment of the two pioneers. But such was their perseverance and enthusiasm that in 1864 Huggins and Miller were able to present ground-breaking results on the spectra of some fifty stars, unambiguously identifying the lines they had found with those in the spectra of terrestrial elements. As Huggins later wrote:

> One important object of this original spectroscopic investigation of the light of the stars and other celestial bodies, namely to discover whether the same chemical elements of those of our earth are present throughout the universe, was most satisfactorily settled in the affirmative; a common chemistry, it was shown, exists through-out the universe.

The new science of astrophysics was on its way.

ABSOLUTELY NEBULOUS . . .

It was late in the summer of 1864 that Huggins made one of his most spectacular discoveries. For more than two centuries, astronomers had debated the nature of the nebulae – those mysterious fuzzy patches in the sky that were neither stars nor planets. Many considered them to be aggregations of stars too distant to be resolved into individual points of light and, indeed, that is what some of them eventually turned out to be. Today, we call those particular specimens galaxies.

But what were the symmetrical objects that Sir William Herschel had christened 'planetary nebulae' in 1785? Herschel himself, having discovered a bright star at the exact centre of one such nebula in 1790, was convinced that they were not made of stars – but could offer no convincing alternative. Then, in 1864, along came Huggins with his spectroscope.

> On the evening of August 29th [he wrote in 1899] I directed the telescope for the first time to a planetary nebula in Draco. I looked into the spectroscope. No spectrum such as I expected! A single bright line only! At first I suspected some displacement of the prism, and that I was looking at a reflection of the illuminated slit from one of its faces. This thought was scarcely more than momentary; then the true interpretation flashed upon me. The riddle of the nebulae was solved. The answer, which had come to us in the light itself, read: Not an aggregation of stars, but a luminous gas.

Huggins quickly followed up this remarkable observation with similar studies of the Great Nebula in Orion and a handful of other diffuse

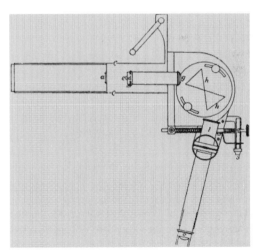

Figure 2. The rudimentary prism spectroscope built by Huggins and Miller in 1862. Light from the telescope enters on the left and, after being made parallel by a collimator lens (g), is dispersed into its component rainbow colours by two prisms (h). The resulting spectrum is then observed with a small telescope (l).

nebulae. In each case he found the same thing – bright emission lines from a luminous gas. The riddle was indeed solved. At least, for the time being.

Thus was the enormous power of Huggins's technique demonstrated. And, with such a major discovery under his belt, he went from strength to strength, observing everything from comets to novae, and, in the process, pioneering the use of photography as a tool in astronomical spectroscopy. He was supported throughout by his wife (herself an accomplished amateur astronomer), and no doubt was as delighted as she was when she found herself transformed from plain Mrs 'Uggins to Lady Margaret on his knighthood in 1897. Huggins became a major figure in the Royal Society, serving as its president from 1900 to 1905. By the time of his death in 1910, the whole world of astronomy had accepted the vital role of spectroscopy in opening up the physics of the stars. As Huggins himself had written as early as 1866:

> So unexpected and important are the results of the application of spectrum analysis to the objects in the heavens, that this method of observation may be said to have created a new and distinct branch of astronomical science.

And he was quite right.

Figure 3. Sir William Huggins with one of his later astronomical spectroscopes. The prism assembly is in the bottom right-hand corner of the photograph, and the eyepiece is near Huggins's left hand.

Figure 4. The Great Nebula in Orion, which was shown by Huggins in 1864 to be a cloud of glowing gas. (Courtesy David Malin, © Anglo-Australian Telescope Board.)

But what of the spectra of nebulae? Well, actually, there was a problem. And rather an embarrassing one. Most of the bright emission lines seen in the spectra of the nebulae could not be identified with terrestrial elements. True, there were lines in the blue and violet part of the spectrum that seemed to coincide with lines known to be emitted by hydrogen, but others didn't fit any known pattern. In particular, the brightest and most prominent lines, which were in the green part of the nebular spectrum, defied explanation. This mystifying lack of correspondence was not something that could be put down to a velocity shift caused by the Doppler effect, for example. The nebular lines simply had no terrestrial counterparts.

What happened next was perfectly logical, and followed a similar conundrum that had emerged in August 1868 during a total eclipse of the Sun. On that occasion, a number of well-known scientists, including the French astronomer Georges Rayet, had made spectroscopic observations of solar prominences – huge clouds of glowing gas billowing from the Sun's surface. Rayet had found no less than nine bright emission lines, among which was one he took to be sodium D, the well-known orange-yellow line that gives sodium street lamps their characteristic colour.

But further investigation by Norman (later Sir Norman) Lockyer and his colleague, Professor Frankland, soon revealed that this was in fact a different spectrum line – and one that did not correspond with any known terrestrial substance. They therefore assumed that it originated in a new element, unknown on Earth, which they eagerly christened 'helium'. This rash act of faith in the power of spectroscopy was vindicated in 1895, when terrestrial helium was extracted from the mineral cleveite by an enterprising chemist called Ramsay, who boiled it in weak sulphuric acid. It is easy to imagine the glee with which Lockyer and Frankland must have greeted the news of this feat of latter-day alchemy.

Astronomers worried by the lack of identification of the nebular lines took heart from the work of Lockyer and Frankland and, following their lead, associated the mysterious green lines in the spectra of nebulae with a hypothetical new element, which they called 'nebulium'. Nice name, you have to admit. And, in the gung-ho climate of the day, it was probably the only sensible thing to do. However, as time went by, more and more laboratory experiments failed to reveal any evidence of the nebulium lines in terrestrial spectra, and gloom settled once more over the nebular camp.

The failure was thrown into stark relief by the increasingly precise wavelengths that were being determined by both laboratory and astronomical spectroscopists. Any room for doubt about the exact position of these lines was rapidly disappearing. Moreover, the periodic table of the elements didn't seem to have any gaps among light elements where you would expect nebulium to appear. So, could it perhaps be something that was extraordinarily rare in the Universe? Absolutely not – the stuff was everywhere. The brilliance of the green lines testified to its abundance in the gaseous nebulae. As the twentieth century dawned, flourished, and was all but obliterated in the carnage of the Great War, the mystery of nebulium deepened into a constant irritation in the minds of the world's physicists.

FORBIDDEN LINES

It was not until 1927 that the first glimmer of light began to shine on the problem. In a book that was still a standard text when your humble author was a lad (albeit by then in a much later edition), a gifted American astronomer threw out an illuminating suggestion. His name was already well known from the colour-magnitude diagram for stars that he and the Danish astronomer Ejnar Hertzsprung had discovered independently in the early years of the twentieth century. Henry Norris Russell was one of the leading lights of the Princeton University Observatory and, when he collaborated with his colleagues Raymond Smith Dugan and John Quincy Stewart to write a textbook entitled simply *Astronomy*, he speculated sagely on the origin of the nebulium lines: '. . . it is now practically certain that they must be due not to atoms of unknown kinds but to atoms of known kinds shining under unfamiliar conditions'.

Russell went further, suggesting that those unfamiliar conditions might be what you would find in a gas of very low density. And he postulated that the mechanism for the emission of such unknown lines would be that it took 'a relatively long time (as atomic events go) for an atom to get into the right state to emit them' – a state that under normal laboratory conditions would be rudely interrupted by a collision with a neighbouring atom. In the extremely low pressure prevailing in a gaseous nebula, atomic collisions would be rare, and who knows what might happen to the energy states of these relatively undisturbed atoms?

The person who answered that question in characteristically brilliant and lucid fashion was the hero of our doggerel abstract, another American called Ira Sprague Bowen. Working at the California Institute of Technology, Bowen had been calculating the possible energy states that can be exhibited by various atoms, and the wavelengths of light emitted when the atoms jump from one energy state to another. Such transitions are, in fact, the origin of all the emission lines observed in the spectra of glowing gases, whether in a flame in the laboratory or a nebula in the depths of space.

Physicists had learned that these energy jumps were governed by certain selection rules, and that some transitions were permitted, while others were forbidden. That rather draconian term was, in fact, a bit misleading, as the 'forbidden' energy jumps were not really forbidden. More accurately, they were highly improbable under laboratory conditions, because the atoms bump each other into new energy states long before the forbidden radiation can be emitted.

Bowen had been pondering these various transitions, while also reading Russell, Dugan and Stewart's *Astronomy*. Suddenly, the penny dropped . . .

> One night, I went down to work and came home about nine o'clock . . . and started to undress [he recounted in 1968]. As I got about half undressed, I got to thinking about what happens if atoms get into one of those states. Are they stuck there forever? Then it occurred to me, having read this [i.e. Russell's comments on nebulium], maybe they can jump if undisturbed in a nebula, but we can't see them here because of collisions . . . Well, I quickly put a reverse on my dressing, and went down to the lab again. Since I had these levels it was very easy to take these differences and check them up . . . it was a matter of minutes to establish it . . . I worked until midnight and I knew I had the answer when I went home that night.

The answer was that the nebulium lines were caused by a forbidden transition between different energy states of oxygen atoms. These particular atoms had been stripped of two of their electrons – a condition known in the trade as 'doubly ionised'. In modern notation, they are lines of [OIII], where the square brackets denote a forbidden transition and the Roman III rather perversely indicates a double ionization (because a Roman I means a neutral, or un-ionized atom).

Bowen's familiarity with the energy states meant that he could quickly narrow down the possibilities, and then calculate accurate wavelengths for the light that would be emitted by these forbidden [OIII] transitions. They agreed exactly with the wavelengths of the nebulium lines. The puzzle that had dogged astrophysicists for more than sixty years was solved – and the answer lay in those shady-sounding forbidden lines. In the flurry of inspired research that followed, Bowen was also able to match other previously unidentified nebular lines with terrestrial elements and, in the process, solve most of the outstanding problems of nebular astronomy. He quickly dashed off a note for the *Publications of the Astronomical Society of the Pacific* and then, in 1928, presented his results in a seminal paper in the *Astrophysical Journal*. It makes pretty exciting reading, even today.

Ira Bowen was in his late twenties when he unravelled the mystery of nebulium, and he went on to a most distinguished career. For eighteen years (1946–64) he was Director of the Mount Wilson and Palomar Observatories, overseeing the completion of both the 200-inch (5-m) Hale Telescope and the 48-inch (1.2-m) Palomar Schmidt (now called the Oschin–Schmidt Telescope). He died in 1973 at the age of seventy-five.

DISPELLING SUPERSTITION?

It has to be admitted that as a postscript to this great detective story, the final stanza of that florid abstract we started with might be exaggerating slightly. There can't be many people out there today who go around thinking that the Universe is full of nebulium. On the other hand, there do seem to be plenty of folk worrying about alien invasions, Moon-landing conspiracies and faces on Mars. Science has to be ever on its guard to keep the facts well up there in the public consciousness.

One of the astronomer's most effective weapons in the fight against ignorance is the same one William Huggins wielded to such effect in the nineteenth century – the spectroscope, and its photographic counterpart, the spectrograph. Things have moved on a bit; today's spectrographs have a level of sophistication that stretches technology to its very limits. They are so frugal with the faint whispers of light from distant objects that almost none is wasted. They no longer use Huggins-style photography to record the spectra, but have special, chilled TV cameras that are almost as efficient as the laws of physics will allow. And some spectrographs have been designed with such inspired brilliance that they are capable of observing not just one target at a time, but many hundreds. In that way, the efficiency of data collection has been improved beyond all recognition in recent years.

For its part, astrophysics is still intoxicated with the wealth of information that comes from optical (visible light) spectroscopy. Even the most basic observations demonstrate that the Universe is expanding, and that unseen planets orbit our neighbouring stars. And spectroscopy allows us to probe the limits of our understanding – in the mysterious dark matter of the Universe, for example.

So there you have it. As a tool for unravelling the innermost secrets of nebulae, or as a harvester of hard facts from the sky, you can't beat a good spectrograph. They are – in the world of astronomy, at least – absolutely fabulous.

FURTHER READING

Bennett, J.A., 1984, *The Celebrated Phaenomena of Colours: the early history of the spectroscope*, Whipple Museum of the History of Science.

Bowen, I.S., 1928, *Astrophysical Journal*, **67**, 1–15.

Hearnshaw, J.B., 1986, *The Analysis of Starlight: one hundred and fifty years of astronomical spectroscopy*, Cambridge University Press.

King, Henry C., 1955, *The History of the Telescope*, Griffin, London.

Lang, Kenneth R., and Gingerich, Owen, 1979, *A Source Book in Astronomy and Astrophysics, 1900–1975*, Harvard University Press.

Russell, Henry N., Dugan, Raymond S., and Stewart, John Q., 1926 (Vol. I) and 1927 (Vol. II), *Astronomy*, Ginn, Boston.

Thackeray, A.D., 1961, *Astronomical Spectroscopy*, Eyre & Spottiswoode, London.

A History of the Transits of Venus

ALLAN CHAPMAN

Only five transits of Venus across the disc of the Sun have been witnessed by human beings. One was observed in the seventeenth century, two in the eighteenth, two in the nineteenth, and none in the twentieth. They are caused by a line-of-sight effect whereby the orbit of Venus, which is inclined by 3° 23½ minutes to that of the Earth, makes Venus appear to pass across the solar disc at a time when the Sun, Venus and the Earth are all in a straight line. Because of the complex dynamics of the Venus–Earth–Sun system, transits occur in pairs, eight years apart, with a gap of either 105 or 122 years between them.

The realization that Venus could indeed transit the Sun came out of that vast treasury of positional astronomical data accumulated by Tycho Brahe before 1601 and reduced and analysed by Johannes Kepler. And while Kepler, unlike Tycho, was a Copernican, believing that the Earth and planets rotated around the Sun, he was able to use Tycho's data to prove that the Earth did indeed move in space. Kepler even calculated that the inner planets, Mercury and Venus, being closer to the Sun than the Earth, must from time to time cross the Earth's line of sight, and appear to transit the Sun. Indeed, he further correctly calculated that *two* transits would occur in 1631. One was a Venus transit on December 6, 1631, though due to its timing it was not visible from Europe which (with the exception of the Jesuits in China) was where *all* the telescopic astronomy was then being done. The other was a much more frequent transit of Mercury, which occurred on November 7, 1631, and which was successfully observed and published by Pierre Gassendi in Paris.

JEREMIAH HORROCKS AND THE TRANSIT OF 1639

Yet Kepler had not realized that Venus transits occurred in pairs, and it was the English amateur, Jeremiah Horrocks of Liverpool, then living at Much Hoole, near Preston, Lancashire, who successfully guessed this possibility. Horrocks came to this conclusion after trying to calculate the orbital elements of Venus's Inferior Conjunction, due take place towards the end of November 1639. The Inferior Conjunction occurs when Venus passes exactly between the Earth and the Sun, crossing its 'nodal' point, the node being the point at which the inclined planes of the terrestrial and Venusian orbits intersect.

To find this nodal point, and the moment at which the Earthly and Venusian orbital planes would intersect, Horrocks made a series of calculations based on the planetary data contained in the ephemerides of Kepler, Boulliau, Lansberg and other contemporary astronomical mathematicians, although most of these men drew their primary observational data from Tycho. Horrocks found that some predicted that Venus would cross the nodal point *above* the Sun, while others said it would pass *below*. The angle, after all, is only a relatively small one, being about half a degree, and within these parameters of error Horrocks reckoned that on November 24, 1639 Venus would pass directly across, or transit, the Sun's disc.

In 1639 Horrocks was only twenty years old, though his surviving correspondence tells us that his serious interest in astronomy, and in particular the *new* astronomy of Copernicus, Tycho, Galileo, Kepler, Gassendi and others, whose works he cites in his letters, went back to his undergraduate days at Emmanuel College, Cambridge, when he was only about fourteen years old. We do not know what Horrocks was doing in Much Hoole in November 1639. He could not have been the curate of the parish, as some have suggested, for at twenty he was too young even to be a deacon, let alone a priest. He may have worked as a tutor to the children of the local Stones family, and could have been the Bible Clerk of St Michael's Church, but almost certainly, had he lived, he would have been ordained, and have embarked upon a career in the church, as most young men from the universities did.

As soon as Horrocks had calculated that a Venus transit really was likely to take place, in late October 1639 he wrote to his brother Jonas, still living in Liverpool, and to his astronomical friend William Crabtree in Salford, alerting them to keep watch. There is no record that Jonas Horrocks saw anything, though Jeremiah Horrocks and Crabtree did see Venus in transit within half an hour of sunset on November 24, 1639 (old Julian Calendar date). As in Gassendi's observation of the 1631 Mercury transit, both Horrocks and Crabtree used their modest refracting telescopes to project an image of the Sun on to 6-inch-diameter circles drawn on a sheet of paper and pinned to pieces of wood.

Indeed, the Sunday on which the transit took place, and for which Horrocks could only keep watch intermittently because of what seemed to be duties in the parish church, was cloudy. But at 3.15 p.m., as it was setting, the Sun broke through the cloud, and both Horrocks and Crabtree, working twenty-five miles apart, saw the jet-black disc of Venus just edging into the solar disc. Within the course of thirty minutes, Horrocks made three measured observations of Venus as the planet moved across the Sun. From these observations, he was able to draw conclusions that would change the way in which astronomers thought of the Solar System.

For one thing, Horrocks was able to discover the angular diameter of Venus at Inferior Conjunction as a fraction of the known angular diameter of the Sun, and it was only one minute, twelve seconds and not the three minutes assumed up until that time. Instead of being large, Venus was probably surrounded by a brilliantly reflective atmosphere, through which the Sun shone during transit, revealing the true size of the solid planet. Horrocks was also able to establish several key values in Venus's orbital relationship with the Sun, and these were confirmed by Crabtree's independent observations in Salford.

Jeremiah Horrocks's account of his and Crabtree's observations, written up in his *Venus in Sole Visa* (*Venus in Transit across the Sun*) was eventually published in Danzig by Johannes Hevelius in 1662. It says something for Horrocks's scientific fame by this date that his works were published in Latin in Poland. For sadly, Horrocks had died suddenly in January 1641, aged twenty-two, while his friend Crabtree had also passed away in 1644, aged only thirty-four. But recently, in the summer of 1995, Horrocks's original manuscript draft of his Venus observations was offered for sale at Sotheby's, London, from a private collection, and was purchased by Cambridge University Library. For Horrocks and Crabtree, and their friend William Gascoigne of Leeds (who did not see the transit of 1639) were Great Britain's first modern astronomers of truly continental standing, who took the ideas of Kepler and Galileo and developed them further.

THE FAR-SIGHTEDNESS OF EDMOND HALLEY

This originality was recognized later in the seventeenth century by John Flamsteed, the first Astronomer Royal, Robert Hooke, Sir Isaac Newton and Dr Edmond Halley. And it was Halley's own observation, at the age of twenty-one, of a Mercury transit across the Sun when on the South Atlantic island of St Helena in 1677 that made him realize the enormous importance of planetary transits as a way of measuring the dimensions of the Solar System. For if two sets of astronomers, let us say in Greenwich and in St Helena respectively, observe either Mercury or Venus in transit, each set will see the transiting planet in a slightly different position on the Sun's disc. This will be the result of a line-of-sight effect, Mercury and Venus being much closer to us than is the Sun.

Now if we can work out the distance between Greenwich and St Helena – which we can from their known latitudes and longitudes – we can use this distance as a base line from which to triangulate the planet. And knowing the distance of the planet, in conjunction with the proportions implicit within Kepler's Law, we can work out the 'solar parallax', from which we can derive the 'Astronomical Unit', or distance of the Sun.

In Halley's time, astronomical measurements based upon other planetary triangulations had led astronomers to think that the Sun was about eighty million miles from the Earth. However, Halley believed that it was essential to determine the elements of our elliptical orbit with much greater precision. The planetary transits method, therefore, seemed the best way forward. Yet while Mercury transits are relatively common, Halley's experiences on St Helena showed him that Mercury was not the best planet to use. For one thing, seen through the telescopes of the time, it was only a tiny black dot, and its motion was too swift to allow critical timings. Halley realized, however, from their 1662 Danzig publication, that both Horrocks and Crabtree had seen Venus as a distinct *disc* rather than a dot in 1639, and that its motion was quite slow. If, therefore, one could observe one limb, or edge, of the Venusian disc first making contact with the Sun, followed by the first appearance of the complete silhouetted planet, time the planet's exact duration of transit with a pendulum clock, and then watch Venus exit the opposite side of the Sun, one would have a very precise set of data. And astronomers in Greenwich, Milan, India, St Helena and Cape Town would all get slightly different results, depending on their stations. And when all of this data was collated internationally, one should be able to extract a value for the solar parallax.

It says a great deal for Halley's thoroughness and sheer far-sightedness that, while he knew that he would never live to see the next transits of June 6, 1761 and June 3, 1769, he not only calculated their various elements, but published major papers in 1691 and 1716 to ensure that astronomers of two or three generations hence would be fully alerted to them. Halley's clarion call would not be forgotten, especially after the less important Mercury transit of 1753 led the French astronomer Joseph Nicolas Delisle to develop Halley's work further, and publish maps that would show the astronomers of Europe and the Americas where in the world the various phases of the transits could best be observed.

THE 1761 TRANSIT AND THE 'BLACK DROP'

By the late 1750s telescopic astronomy was being practised all across Europe, Western Russia, the Eastern seaboard of the North American continent, the Caribbean and parts of Spanish Central and South America, and most western countries had firmly established scientific academies. The most active of these were the Royal Society of London and the French Académie, and both were assiduous in planning transit observations, not only in Europe but across their growing overseas empires. The only problem was that in 1761 both France and England and their allies were locked in the global conflict of the 'Seven Years War', which raged not only across Europe but also in India, Canada and the Caribbean. And while, in accordance with the gentlemanly conventions of eighteenth-century warfare, both the British and the French governments instructed their respective sets of sea captains not to attack enemy vessels that were on transit expeditions, this practice became unworkable on the high seas.

The British vessel *HMS Seahorse* that was carrying Charles Mason and Jeremiah Dixon to observe the transit in Bencoolen, India, had to limp back home to Plymouth with eleven dead, after an encounter with a larger French man-o'-war. And Captain Fletcher, of *HMS Plassey*, doubted the validity of Alexandre-Gui Pingré's British-granted safe conduct when *Plassey* captured his ship in the Indian Ocean. The French Académicien J.-B. Chappe D'Auteroche had to make a colossal detour around the war zones of eastern Europe even before beginning his long transcontinental haul to observe the transit at Tobolsk in Siberia. And it was also the 1761 transit which gave rise to one of the very first North American astronomical expeditions, when John Winthrop of Harvard College headed a group of Boston scientists bound for Canada to observe the event from a high north-westerly latitude.

In total, some 120 scientists saw the transit of June 6, 1761, from some sixty-two separate locations that spanned not only the whole of Europe but also North America, Canada, Siberia, South Africa and India. The purpose behind this massive geographical spread, of course, was to obtain the maximum angular displacements and longest geographical base lines from which it was hoped that the most precise parallax angle of Venus on the Sun's disc could be obtained.

Yet when all of these results had been collated, the value for the solar parallax for the different observing stations still sprawled between 8.28 and 10.60 arc seconds, which was not as exact as expected. The discrepancy arose not from any imperfection in instruments or observation, but because of a totally unexpected phenomenon which came to be called the 'black drop'. Crucial to an accurate observation of the transit was an exact set of timings for the precise moments when Venus made and then broke contact with the solar limb. But the 'black drop' was a black, blurred filament which appeared whenever Venus approached the solar limb, and made the exact moment of contact hard to establish. Astronomers were right when they surmised that this 'black drop' was the product of the Sun's light being refracted and distorted by the Venusian atmosphere.

By the time astronomers set sail on their expeditions to observe the 1769 transit, not only were they alerted to the 'black drop' and were trying to find ways of compensating for it, but the world was at peace. No expeditions were involved in Anglo-French frigate dogfights on the high seas, and the gentlemen of both nations were treating each other with their usual peacetime cooperation.

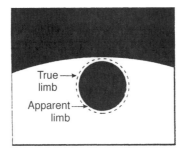

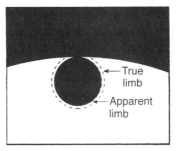

Figure 1. Observers found it extremely difficult to obtain accurate transit timings for Venus because of the so-called 'black drop' effect. As Venus passes on to the Sun's disc, it seems to draw a strip of blackness after it, and when this disappears the transit has already begun. Of course, the effect is due solely to Venus's atmosphere, so there is very little that can be done about it.

THE 1769 TRANSIT

No less than 151 astronomers observed the June 3, 1769 Venus transit from seventy-seven stations worldwide. One of the most far-reaching, in global history terms, was the expedition commanded by Captain James Cook RN to Tahiti (to form an Antipodean observing station halfway around the world from Europe) which would later go on to search for *Terra Incognita Australis* in the Great South Sea, and thereby discover Western Australia. Chappe D'Auteroche, who had struggled with warring armies, packs of starving wolves and bitter cold when he went to Siberia in 1761, was now sent by the French Académie to the then wilds of California via the overland route through the steaming jungles of Central America. But after successfully observing the transit, Chappe D'Auteroche and his companions died of tropical fever, leaving the expedition's engineer Monsieur Pauli as the sole survivor, who got the valuable results back to Europe.

Yet no one can discuss the history of the 1761 and 1769 Venus transits without being amazed at the courage, determination and sheer bad luck of Guillaume-Joseph-Hyacinthe-Jean-Baptiste Le Gentil, who had been dispatched by the Académie to observe the 1761 transit at the French outpost at Pondicherry, India. By the time of his arrival, however, the fortunes of war had put Pondicherry in British hands, and his ship narrowly avoided capture or sinking. And on the day of the transit, although the Sun beamed down from a cloudless sky, it was impossible to make an observation from the heaving deck of a ship in the Indian Ocean.

Not to be deterred, Le Gentil did not go home to France, but spent the next eight years in the lands that surrounded the Indian Ocean, conducting a remarkable study into the natural historical, oceanographic, and ethnological treasures of the Far East. And when the world was firmly at peace in 1769, he went back to Pondicherry to make his transit observation. Here he was given the warmest welcome by the British Governor, Mr Law, who was a gentleman of culture, and whose company Le Gentil enjoyed. Not only did Mr Law intend to act as Le Gentil's assistant for the transit observation, but he even obtained the loan of an excellent new Dollond achromatic refractor – a telescope much better than Le Gentil's own instrument – with which to make the observation.

June 3, 1769 dawned brilliantly clear. Yet at the crucial moment, when Venus made its first contact with the Sun, a flimsy cloud passed across the disc. Le Gentil had missed his second transit. On his return to France from his long absence, he found himself presumed dead, and his relatives carving up his estates. Then came the French Revolution which occasioned him to flee to England, where he died at the beginning of the nineteenth century.

The combined international observations for the 1769 transit, however, gave a much more consistent value for the solar parallax than those of 1761, falling as it did between the error spread 8.43 to 8.80 arc seconds, which is pretty close to the present-day figure of 8.794148. When all the results were compared, late-eighteenth-century astronomers were able to ascribe dimensions to the Solar System which are very close to those which we accept today, though in 1824 the German astronomer Johann Encke's masterly reduction of the eighteenth-century transit observations resulted in an Astronomical Unit of 95,000,000 miles rather than the 93,000,000 which we accept nowadays as the mean distance.

THE NINETEENTH-CENTURY TRANSITS

Nineteenth-century astronomy was dominated by precision, and astronomers worldwide awaited the Venus transits of 1874 and 1882. Sir George Biddell Airy, the Astronomer Royal, virtually took charge of the preparations for the 1874 transit, arranging a variety of global expeditions, and ensuring the training of observers in how to do such things as correct for the 'black drop'. By 1874, however, there were per-manent and operational observatories all across North America, South Africa, India, Australasia, Japan and the Russian Empire, in a world that had already been shrunk by the steamship, the railway and the electric telegraph. The global observation of the transit, therefore, was going to be vastly more complete than in 1769. Consequently the expeditions that actually were sent out went for the most part to remote islands in vast oceans to make the observational coverage as thorough as possible. A British expedition, for instance, was sent to Hawaii in the Pacific in 1874, while Father Stephen Perry FRS, the Jesuit Director of the Stonyhurst College Astrophysical Observatory in Lancashire, took an expedition to the uninhabited and windswept Kerguelan Island in 1874 to cover a remote station in the southern Indian Ocean. In 1882 Father Perry led another transit expedition to a fever-ridden area of Madagascar.

One of the best-equipped expeditions that set out to observe the transit of December 9, 1874 was that which was entirely paid for out of the private resources of the Scottish grand amateur astronomer Lord Lindsay, Earl of Crawford and Balcarres, which went to the island of Mauritius. Lord Lindsay, moreover, like most of the other transit astronomers by 1874, used photography to record the position of Venus on the solar disc, and this, combined with the colossal improve-ments which had taken place in astronomical instrument-making since 1769, meant that after 1882 the Astronomical Unit was known with modern precision.

But one of the most remarkable observers who photographed a Venus transit – in this case, that of December 6, 1882 – came from the opposite end of the social scale from Lord Lindsay, and yet clearly pursued astronomy with an equal passion. This was Samuel Cooper, a bricklayer who lived in the Dorset village of Charminster. Samuel Cooper had built a 9-inch-aperture silver-on-glass reflecting telescope, and fitted it with a plateholder, making it a powerful astronomical camera. On a winter's day of poor weather conditions that was reminiscent of that on which Horrocks and Crabtree observed the first transit of 1639, Cooper obtained three excellent photographs of Venus as it moved across the Sun, taken at 2.15, 2.30, and 2.45 p.m. respectively. And Samuel Cooper's dedication in the building and using of his reflector indicates how widely a serious interest in astronomy had dispersed throughout the population of Britain by 1882.

THE FORTHCOMING VENUS TRANSITS

By the time the next transits take place in 2004 and 2012, they will have long since lost the scientific urgency of their predecessors. It is now possible to measure seasonal changes in the Astronomical Unit to astonishing levels of accuracy almost wherever we wish without the need to wait for a planetary transit. Similarly, space probes have mapped the surface and other physical characteristics of Venus to a level of detail that no astronomer of 1882 could have possibly imagined.

On the other hand, the forthcoming Venus transits will be well worth observing. Not only will the transits be beautiful sights to behold in their own right, but they will also be great reminders of the sheer doggedness and determination with which astronomers of previous centuries pursued their science. And if you choose to combine your observation of the transits with a holiday in an exotic location, just remember that your predecessors in 1761 and 1769 had been willing to risk shipwreck, war, fever, starving wolves, Arctic cold, steaming jungles and unsympathetic locals to observe Venus in transit in far-flung regions of the globe.

FURTHER READING

Chapman, Allan, 'Jeremiah Horrocks, the transit of Venus and the "New Astronomy" in seventeenth-century England', *Quarterly Journal of the Royal Astronomical Society* 31 (1990), pp. 333–57. Reprinted in Chapman, *Astronomical Instruments and their Users* (Variorum, Aldershot, 1996) V.

Chappe D'Auteroche, J.-B., 'Extract from a Journey to Siberia, for Observing the Transit of Venus over the Sun: by M. L'Abbé Chappe d'Auteroche', *Gentlemans' Magazine*, xxxiii (1763) pp. 547–552.

Chauvin, Michael E., *Hokulo'a. The British 1874 Transit of Venus Expedition to Hawaii* (Bishop Museum Press, Hawaii, 2003). I thank Dr Chauvin for providing me with so much information about the 1874 transit preparation and observation, and for graciously giving me a pre-publication copy of *Hokulo'a.*

Horrocks (Horrox), Jeremiah, *The Transit of Venus across the Sun* [translation of *Venus in Sole Visa*], translated by A.B. Whatton and prefaced by Whatton's *Memoir of the Life and Labours of the Rev. Jeremiah Horrocks* (London, 1859).

Nunis, Doyce Blackman, *The 1769 Transit of Venus, the Baja California Observations of Jean-Baptiste Chappe D'Auteroche, Vincente de Doz, and Joaquin Velazque Cardenas de Leon* [being a surviving narrative from Chappe D'Auteroche's California journey] (Natural History Museum, Los Angeles County, 1982).

Woolf, Harry H., *The Transits of Venus: A Study of Eighteenth-Century Science* (Princeton University Press, 1959).

Part III

Miscellaneous

Some Interesting Variable Stars

JOHN ISLES

All variable stars are of potential interest, and hundreds of them can be observed with the slightest optical aid – even with a pair of binoculars. The stars in the list that follows include many that are popular with amateur observers, as well as some less well-known objects that are nevertheless suitable for study visually. The periods and ranges of many variables are not constant from one cycle to another, and some are completely irregular.

Finder charts are given after the list for those stars marked with an asterisk. These charts are adapted with permission from those issued by the Variable Star Section of the British Astronomical Association. Apart from the eclipsing variables and others in which the light changes are purely a geometrical effect, variable stars can be divided broadly into two classes: the pulsating stars, and the eruptive or cataclysmic variables.

Mira (Omicron Ceti) is the best-known member of the long-period subclass of pulsating red giant stars. The chart is suitable for use in estimating the magnitude of Mira when it reaches naked-eye brightness – typically from about a month before the predicted date of maximum until two or three months after maximum. Predictions for Mira and other stars of its class follow the section of finder charts.

The semi-regular variables are less predictable, and generally have smaller ranges. V Canum Venaticorum is one of the more reliable ones, with steady oscillations in a six-month cycle. Z Ursae Majoris, easily found with binoculars near Delta, has a large range, and often shows double maxima owing to the presence of multiple periodicities in its light changes. The chart for Z is also suitable for observing another semi-regular star, RY Ursae Majoris. These semi-regular stars are mostly red giants or supergiants.

The RV Tauri stars are of earlier spectral class than the semi-regulars, and in a full cycle of variation they often show deep minima and double maxima that are separated by a secondary minimum. U Monocerotis is one of the brightest RV Tauri stars.

Among eruptive variable stars is the carbon-rich supergiant R Coronae Borealis. Its unpredictable eruptions cause it not to brighten, but to fade. This happens when one of the sooty clouds that the star throws out from time to time happens to come in our direction and blots out most of the star's light from our view. Much of the time R Coronae is bright enough to be seen with binoculars, and the chart can be used to estimate its magnitude. During the deepest minima, however, the star needs a telescope of 25 centimetres or larger aperture to be detected.

CH Cygni is a symbiotic star – that is, a close binary comprising a red giant and a hot dwarf star that interact physically, giving rise to out-bursts. The system also shows semi-regular oscillations, and sudden fades and rises that may be connected with eclipses.

Observers can follow the changes of these variable stars by using the comparison stars whose magnitudes are given below each chart. Observations of variable stars by amateurs are of scientific value, provided they are collected and made available for analysis. This is done by several organizations, including the British Astronomical Association (see the list of astronomical societies in this volume), the American Association of Variable Star Observers (25 Birch Street, Cambridge, Mass. 02138), the Royal Astronomical Society of New Zealand (P.O. Box 3181, Wellington), and the Internet group VSNET (http:vsnet.kusastro.kyoto-u.ac.jp/vsnet/index.html).

Star	RA		Declination		Range	Type	Period	Spectrum
	h	m	°	′			(days)	
R Andromedae	00	24.0	+38	35	5.8–14.9	Mira	409	S
W Andromedae	02	17.6	+44	18	6.7–14.6	Mira	396	S
U Antliae	10	35.2	−39	34	5–6	Irregular	—	C
Theta Apodis	14	05.3	−76	48	5–7	Semi-regular	119	M
R Aquarii	23	43.8	−15	17	5.8–12.4	Symbiotic	387	M+Pec
T Aquarii	20	49.9	−05	09	7.2–14.2	Mira	202	M
R Aquilae	19	06.4	+08	14	5.5–12.0	Mira	284	M
V Aquilae	19	04.4	−05	41	6.6–8.4	Semi-regular	353	C
Eta Aquilae	19	52.5	+01	00	3.5–4.4	Cepheid	7.2	F–G
U Arae	17	53.6	−51	41	7.7–14.1	Mira	225	M
R Arietis	02	16.1	+25	03	7.4–13.7	Mira	187	M
U Arietis	03	11.0	+14	48	7.2–15.2	Mira	371	M
R Aurigae	05	17.3	+53	35	6.7–13.9	Mira	458	M
Epsilon Aurigae	05	02.0	+43	49	2.9–3.8	Algol	9892	F+B

Star	RA		Declination		Range	Type	Period	Spectrum
	h	m	°	′			(days)	
R Boötis	14	37.2	+26	44	6.2–13.1	Mira	223	M
X Camelopardalis	04	45.7	+75	06	7.4–14.2	Mira	144	K–M
R Cancri	08	16.6	+11	44	6.1–11.8	Mira	362	M
X Cancri	08	55.4	+17	14	5.6–7.5	Semi-regular	195?	C
R Canis Majoris	07	19.5	−16	24	5.7–6.3	Algol	1.1	F
VY Canis Majoris	07	23.0	−25	46	6.5–9.6	Unique	—	M
S Canis Minoris	07	32.7	+08	19	6.6–13.2	Mira	333	M
R Canum Ven.	13	49.0	+39	33	6.5–12.9	Mira	329	M
*V Canum Ven.	13	19.5	+45	32	6.5–8.6	Semi-regular	192	M
R Carinae	09	32.2	−62	47	3.9–10.5	Mira	309	M
S Carinae	10	09.4	−61	33	4.5–9.9	Mira	149	K–M
l Carinae	09	45.2	−62	30	3.3–4.2	Cepheid	35.5	F–K
Eta Carinae	10	45.1	−59	41	−0.8–7.9	Irregular	—	Pec
R Cassiopeiae	23	58.4	+51	24	4.7–13.5	Mira	430	M
S Cassiopeiae	01	19.7	+72	37	7.9–16.1	Mira	612	S
W Cassiopeiae	00	54.9	+58	34	7.8–12.5	Mira	406	C
Gamma Cas.	00	56.7	+60	43	1.6–3.0	Gamma Cas.	—	B
Rho Cassiopeiae	23	54.4	+57	30	4.1–6.2	Semi-regular	—	F–K
R Centauri	14	16.6	−59	55	5.3–11.8	Mira	546	M
S Centauri	12	24.6	−49	26	7–8	Semi-regular	65	C
T Centauri	13	41.8	−33	36	5.5–9.0	Semi-regular	90	K–M
S Cephei	21	35.2	+78	37	7.4–12.9	Mira	487	C
T Cephei	21	09.5	+68	29	5.2–11.3	Mira	388	M
Delta Cephei	22	29.2	+58	25	3.5–4.4	Cepheid	5.4	F–G
Mu Cephei	21	43.5	+58	47	3.4–5.1	Semi-regular	730	M
U Ceti	02	33.7	−13	09	6.8–13.4	Mira	235	M
W Ceti	00	02.1	−14	41	7.1–14.8	Mira	351	S
*Omicron Ceti	02	19.3	−02	59	2.0–10.1	Mira	332	M
R Chamaeleontis	08	21.8	−76	21	7.5–14.2	Mira	335	M
T Columbae	05	19.3	−33	42	6.6–12.7	Mira	226	M
R Comae Ber.	12	04.3	+18	47	7.1–14.6	Mira	363	M
*R Coronae Bor.	15	48.6	+28	09	5.7–14.8	R Coronae Bor.	—	C
S Coronae Bor.	15	21.4	+31	22	5.8–14.1	Mira	360	M
T Coronae Bor.	15	59.6	+25	55	2.0–10.8	Recurrent nova	—	M+Pec
V Coronae Bor.	15	49.5	+39	34	6.9–12.6	Mira	358	C
W Coronae Bor.	16	15.4	+37	48	7.8–14.3	Mira	238	M
R Corvi	12	19.6	−19	15	6.7–14.4	Mira	317	M
R Crucis	12	23.6	−61	38	6.4–7.2	Cepheid	5.8	F–G
R Cygni	19	36.8	+50	12	6.1–14.4	Mira	426	S
U Cygni	20	19.6	+47	54	5.9–12.1	Mira	463	C

Star	RA		Declination		Range	Type	Period	Spectrum
	h	m	°	′			(days)	
W Cygni	21	36.0	+45	22	5.0–7.6	Semi-regular	131	M
RT Cygni	19	43.6	+48	47	6.0–13.1	Mira	190	M
SS Cygni	21	42.7	+43	35	7.7–12.4	Dwarf nova	50±	K+Pec
*CH Cygni	19	24.5	+50	14	5.6–9.0	Symbiotic	—	M+B
Chi Cygni	19	50.6	+32	55	3.3–14.2	Mira	408	S
R Delphini	20	14.9	+09	05	7.6–13.8	Mira	285	M
U Delphini	20	45.5	+18	05	5.6–7.5	Semi-regular	110?	M
EU Delphini	20	37.9	+18	16	5.8–6.9	Semi-regular	60	M
Beta Doradûs	05	33.6	−62	29	3.5–4.1	Cepheid	9.8	F–G
R Draconis	16	32.7	+66	45	6.7–13.2	Mira	246	M
T Eridani	03	55.2	−24	02	7.2–13.2	Mira	252	M
R Fornacis	02	29.3	−26	06	7.5–13.0	Mira	389	C
R Geminorum	07	07.4	+22	42	6.0–14.0	Mira	370	S
U Geminorum	07	55.1	+22	00	8.2–14.9	Dwarf nova	105±	Pec+M
Zeta Geminorum	07	04.1	+20	34	3.6–4.2	Cepheid	10.2	F–G
Eta Geminorum	06	14.9	+22	30	3.2–3.9	Semi-regular	233	M
S Gruis	22	26.1	−48	26	6.0–15.0	Mira	402	M
S Herculis	16	51.9	+14	56	6.4–13.8	Mira	307	M
U Herculis	16	25.8	+18	54	6.4–13.4	Mira	406	M
Alpha Herculis	17	14.6	+14	23	2.7–4.0	Semi-regular	—	M
68, u Herculis	17	17.3	+33	06	4.7–5.4	Algol	2.1	B+B
R Horologii	02	53.9	−49	53	4.7–14.3	Mira	408	M
U Horologii	03	52.8	−45	50	6–14	Mira	348	M
R Hydrae	13	29.7	−23	17	3.5–10.9	Mira	389	M
U Hydrae	10	37.6	−13	23	4.3–6.5	Semi-regular	450?	C
VW Hydri	04	09.1	−71	18	8.4–14.4	Dwarf nova	27±	Pec
R Leonis	09	47.6	+11	26	4.4–11.3	Mira	310	M
R Leonis Minoris	09	45.6	+34	31	6.3–13.2	Mira	372	M
R Leporis	04	59.6	−14	48	5.5–11.7	Mira	427	C
Y Librae	15	11.7	−06	01	7.6–14.7	Mira	276	M
RS Librae	15	24.3	−22	55	7.0–13.0	Mira	218	M
Delta Librae	15	01.0	−08	31	4.9–5.9	Algol	2.3	A
R Lyncis	07	01.3	+55	20	7.2–14.3	Mira	379	S
R Lyrae	18	55.3	+43	57	3.9–5.0	Semi-regular	46?	M
RR Lyrae	19	25.5	+42	47	7.1–8.1	RR Lyrae	0.6	A–F
Beta Lyrae	18	50.1	+33	22	3.3–4.4	Eclipsing	12.9	B
U Microscopii	20	29.2	−40	25	7.0–14.4	Mira	334	M
*U Monocerotis	07	30.8	−09	47	5.9–7.8	RV Tauri	91	F–K
V Monocerotis	06	22.7	−02	12	6.0–13.9	Mira	340	M
R Normae	15	36.0	−49	30	6.5–13.9	Mira	508	M

Star	RA h	m	Declination °	′	Range	Type	Period (days)	Spectrum
T Normae	15	44.1	−54	59	6.2−13.6	Mira	241	M
R Octantis	05	26.1	−86	23	6.3−13.2	Mira	405	M
S Octantis	18	08.7	−86	48	7.2−14.0	Mira	259	M
V Ophiuchi	16	26.7	−12	26	7.3−11.6	Mira	297	C
X Ophiuchi	18	38.3	+08	50	5.9−9.2	Mira	329	M
RS Ophiuchi	17	50.2	−06	43	4.3−12.5	Recurrent nova	—	OB+M
U Orionis	05	55.8	+20	10	4.8−13.0	Mira	368	M
W Orionis	05	05.4	+01	11	5.9−7.7	Semi-regular	212	C
Alpha Orionis	05	55.2	+07	24	0.0−1.3	Semi-regular	2335	M
S Pavonis	19	55.2	−59	12	6.6−10.4	Semi-regular	381	M
Kappa Pavonis	18	56.9	−67	14	3.9−4.8	Cepheid	9.1	G
R Pegasi	23	06.8	+10	33	6.9−13.8	Mira	378	M
X Persei	03	55.4	+31	03	6.0−7.0	Gamma Cas.	—	O9.5
Beta Persei	03	08.2	+40	57	2.1−3.4	Algol	2.9	B
Zeta Phoenicis	01	08.4	−55	15	3.9−4.4	Algol	1.7	B+B
R Pictoris	04	46.2	−49	15	6.4−10.1	Semi-regular	171	M
RS Puppis	08	13.1	−34	35	6.5−7.7	Cepheid	41.4	F−G
L² Puppis	07	13.5	−44	39	2.6−6.2	Semi-regular	141	M
T Pyxidis	09	04.7	−32	23	6.5−15.3	Recurrent nova	7000±	Pec
U Sagittae	19	18.8	+19	37	6.5−9.3	Algol	3.4	B+G
WZ Sagittae	20	07.6	+17	42	7.0−15.5	Dwarf nova	1900±	A
R Sagittarii	19	16.7	−19	18	6.7−12.8	Mira	270	M
RR Sagittarii	19	55.9	−29	11	5.4−14.0	Mira	336	M
RT Sagittarii	20	17.7	−39	07	6.0−14.1	Mira	306	M
RU Sagittarii	19	58.7	−41	51	6.0−13.8	Mira	240	M
RY Sagittarii	19	16.5	−33	31	5.8−14.0	R Coronae Bor.	—	G
RR Scorpii	16	56.6	−30	35	5.0−12.4	Mira	281	M
RS Scorpii	16	55.6	−45	06	6.2−13.0	Mira	320	M
RT Scorpii	17	03.5	−36	55	7.0−15.2	Mira	449	S
Delta Scorpii	16	00.3	−22	37	1.6−2.3	Irregular	—	B
S Sculptoris	00	15.4	−32	03	5.5−13.6	Mira	363	M
R Scuti	18	47.5	−05	42	4.2−8.6	RV Tauri	146	G−K
R Serpentis	15	50.7	+15	08	5.2−14.4	Mira	356	M
S Serpentis	15	21.7	+14	19	7.0−14.1	Mira	372	M
T Tauri	04	22.0	+19	32	9.3−13.5	T Tauri	—	F−K
SU Tauri	05	49.1	+19	04	9.1−16.9	R Coronae Bor.	—	G
Lambda Tauri	04	00.7	+12	29	3.4−3.9	Algol	4.0	B+A
R Trianguli	02	37.0	+34	16	5.4−12.6	Mira	267	M
R Ursae Majoris	10	44.6	+68	47	6.5−13.7	Mira	302	M
T Ursae Majoris	12	36.4	+59	29	6.6−13.5	Mira	257	M

Star	RA		Declination		Range	Type	Period	Spectrum
	h	m	°	'			(days)	
*Z Ursae Majoris	11	56.5	+57	52	6.2–9.4	Semi-regular	196	M
*RY Ursae Majoris	12	20.5	+61	19	6.7–8.3	Semi-regular	310?	M
U Ursae Minoris	14	17.3	+66	48	7.1–13.0	Mira	331	M
R Virginis	12	38.5	+06	59	6.1–12.1	Mira	146	M
S Virginis	13	33.0	−07	12	6.3–13.2	Mira	375	M
SS Virginis	12	25.3	+00	48	6.0–9.6	Semi-regular	364	C
R Vulpeculae	21	04.4	+23	49	7.0–14.3	Mira	137	M
Z Vulpeculae	19	21.7	+25	34	7.3–8.9	Algol	2.5	B+A

V CANUM VENATICORUM   13ᴴ 19.5ᴹ +45° 32′ (2000)

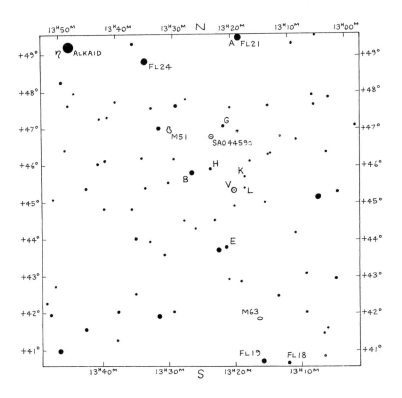

A 5.1	H 7.8
B 5.9	K 8.4
E 6.5	L 8.6
G 7.1	

° (MIRA) CETI   02ᴴ 19.3ᴹ −02° 59′ (2000)

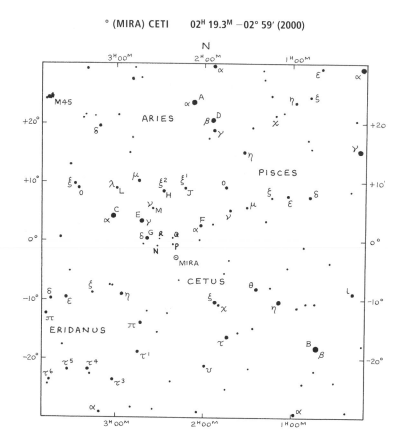

A 2.2	J 4.4
B 2.4	L 4.9
C 2.7	M 5.1
D 3.0	N 5.4
E 3.6	P 5.5
F 3.8	Q 5.7
G 4.1	R 6.1
H 4.3	

R CORONAE BOREALIS   15h 48m 34.4s +28° 09′ 24′′ (2000)

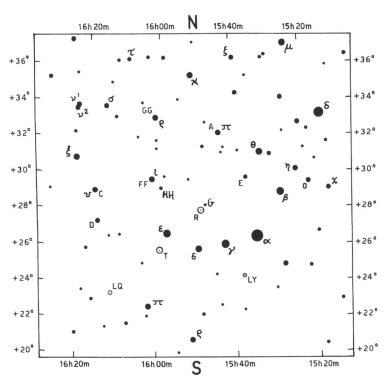

FF 5.0   C 5.8
GG 5.4   D 6.2
A 5.6   E 6.5
         HH 7.1
         G 7.4

CH CYGNI   19^H 24^M 33^S +50° 14.5' (2000)

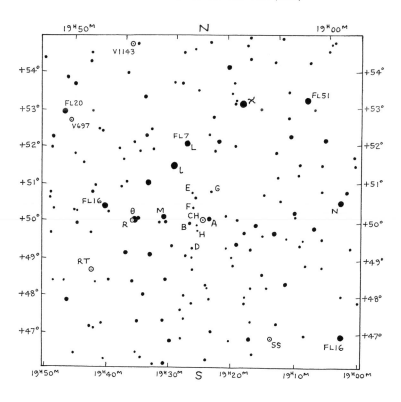

N 5.4	D 8.0
M 5.5	E 8.1
L 5.8	F 8.5
A 6.5	G 8.5
B 7.4	H 9.2

U MONOCEROTIS   07ᴴ 30.8ᴹ 33ˢ −09° 47′ (2000)

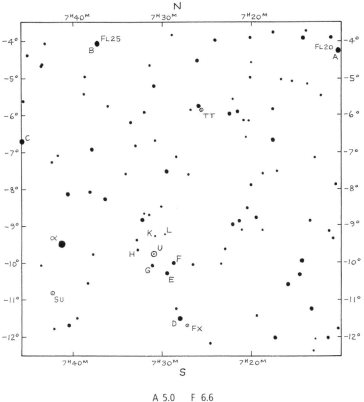

A 5.0	F 6.6
B 5.2	G 7.0
C 5.7	H 7.5
D 5.9	K 7.8
E 6.0	L 8.0

RY URSAE MAJORIS   12^H 20.5^M +61° 19′ (2000)
Z URSAE MAJORIS   11^H 56.5^M +57° 52′ (2000)

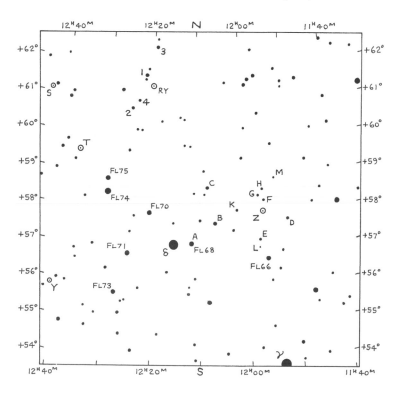

A	6.5	F	8.6	M	9.1
B	7.2	G	8.7	1	6.9
C	7.6	H	8.8	2	7.4
D	8.0	K	8.9	3	7.7
E	8.3	L	9.0	4	7.8

Mira Stars: Maxima, 2004

JOHN ISLES

Below are the predicted dates of maxima for Mira stars that reach magnitude 7.5 or brighter at an average maximum. Individual maxima can in some cases be brighter or fainter than average by a magnitude or more, and all dates are only approximate. The positions, extreme ranges and mean periods of these stars can be found in the preceding list of interesting variable stars.

Star	Mean Magnitude at Maximum	Dates of Maxima
W Andromedae	7.4	Jan 14
R Aquarii	6.5	Sep 5
R Aquilae	6.1	Aug 9
UV Aurigae	7.0	Sep 23
R Bootis	7.2	Jan 20, Aug 30
R Cancri	6.8	Oct 22
S Canis Minoris	7.5	Jul 10
R Carinae	4.6	Mar 21
S Carinae	5.7	Feb 10, Jul 9, Dec 5
R Cassiopeiae	7.0	Mar 25
R Centauri	5.8	May 4
T Cephei	6.0	Oct 13
U Ceti	7.5	Jun 16
Omicron Ceti	3.4	May 30
T Columbae	7.5	Feb 8, Sep 21
S Coronae Borealis	7.3	Oct 12
V Coronae Borealis	7.5	Mar 31
R Corvi	7.5	Sep 11
R Cygni	7.5	Apr 2
RT Cygni	7.3	May 18, Nov 24
Chi Cygni	5.2	May 28
R Geminorum	7.1	Nov 30

Star	Mean Magnitude at Maximum	Dates of Maxima
U Herculis	7.5	May 10
R Horologii	6.0	Jul 12
R Hydrae	4.5	Sep 21
R Leonis	5.8	Aug 5
R Leonis Minoris	7.1	Nov 5
RS Librae	7.5	Feb 25, Sep 30
V Monocerotis	7.0	Nov 20
R Normae	7.2	Nov 13
T Normae	7.4	Mar 28, Nov 24
V Ophiuchi	7.5	May 1
X Ophiuchi	6.8	Apr 23
U Orionis	6.3	Dec 23
R Sagittarii	7.3	Jan 25, Oct 21
RR Sagittarii	6.8	Jun 7
RT Sagittarii	7.0	May 3
RU Sagittarii	7.2	Jul 3
RR Scorpii	5.9	Jul 29
RS Scorpii	7.0	Nov 9
S Sculptoris	6.7	Dec 6
R Serpentis	6.9	Dec 3
R Trianguli	6.2	Aug 8
R Ursae Majoris	7.5	Apr 13
R Virginis	6.9	Jan 9, Jun 3, Oct 26
S Virginis	7.0	Jan 12

Some Interesting Double Stars

R.W. ARGYLE

The positions, angles and separations given below correspond to epoch 2004.0.

No.	RA		Declin-ation		Star	Magni-tudes	Separa-tion	PA	Cata-logue	Comments
	h	m	°	′			arcsec	°		
1	00	31.5	−62	58	β Tuc	4.4, 4.8	27.1	169	LCL 119	Both again difficult doubles.
2	00	49.1	+57	49	η Cas	3.4, 7.5	13.0	319	Σ60	Easy. Creamy, bluish.
3	00	55.0	+23	38	36 And	6.0, 6.4	1.0	314	Σ73	P = 168 years. Both yellow. Slowly opening.
4	01	13.7	+07	35	ζ Psc	5.6, 6.5	23.1	63	Σ100	Yellow, reddish-white.
5	01	39.8	−56	12	p Eri	5.8, 5.8	11.6	190	Δ5	Period = 483 years.
6	01	53.5	+19	18	γ Ari	4.8, 4.8	7.5	1	Σ180	Very easy. Both white.
7	02	02.0	+02	46	α Psc	4.2, 5.1	1.8	269	Σ202	Binary, period = 933 years.
8	02	03.9	+42	20	γ And	2.3, 5.0	9.6	63	Σ205	Yellow, blue. Relatively fixed.
					γ2 And	5.1, 6.3	0.4	103	OΣ38	BC. Needs 30 cm. Closing.
9	02	29.1	+67	24	ι Cas AB	4.9, 6.9	2.6	230	Σ262	AB is long-period binary. P = 620 years.
					ι Cas AC	4.9, 8.4	7.2	118		
10	02	33.8	−28	14	ω For	5.0, 7.7	10.8	245	HJ 3506	Common proper motion.
11	02	43.3	+03	14	γ Cet	3.5, 7.3	2.6	298	Σ299	Not too easy.

No.	RA h	RA m	Declin- ation °	Declin- ation ′	Star	Magni- tudes	Separa- tion arcsec	PA °	Cata- logue	Comments
12	02	58.3	−40	18	θ Eri	3.4, 4.5	8.3	90	PZ 2	Both white.
13	02	59.2	+21	20	ε Ari	5.2, 5.5	1.5	208	Σ333	Binary. Little motion. Both white.
14	03	00.9	+52	21	Σ331 Per	5.3, 6.7	12.0	85	−	Fixed.
15	03	12.1	−28	59	α For	4.0, 7.0	5.0	299	HJ 3555	P = 269 years. B variable?
16	03	48.6	−37	37	f Eri	4.8, 5.3	8.2	215	Δ16	Pale yellow. Fixed.
17	03	54.3	−02	57	32 Eri	4.8, 6.1	6.9	348	Σ470	Fixed.
18	04	32.0	+53	55	1 Cam	5.7, 6.8	10.3	308	Σ550	Fixed.
19	04	50.9	−53	28	ι Pic	5.6, 6.4	12.4	58	Δ18	Good object for small apertures. Fixed.
20	05	13.2	−12	56	κ Lep	4.5, 7.4	2.2	357	Σ661	Visible in 7.5 cm.
21	05	14.5	−08	12	β Ori	0.1, 6.8	9.5	204	Σ668	Companion once thought to be close double.
22	05	21.8	−24	46	41 Lep	5.4, 6.6	3.4	93	HJ 3752	Deep-yellow pair in a rich field.
23	05	24.5	−02	24	η Ori	3.8, 4.8	1.7	78	DA 5	Slow-moving binary.
24	05	35.1	+09	56	λ Ori	3.6, 5.5	4.3	44	Σ738	Fixed.
25	05	35.3	−05	23	θ Ori AB	6.7, 7.9	8.6	32	Σ748	Trapezium in M42.
					θ Ori CD	5.1, 6.7	13.4	61		
26	05	38.7	−02	36	σ Ori AC	4.0, 10.3	11.4	238	Σ762	Quintuple. A is a close double.
					σ Ori ED	6.5, 7.5	30.1	231		
27	05	40.7	−01	57	ζ Ori	1.9, 4.0	2.4	164	Σ774	Can be split in 7.5 cm. Long-period binary.
28	06	14.9	+22	30	η Gem	var, 6.5	1.6	255	β1008	Well seen with 20 cm. Primary orange.
29	06	46.2	+59	27	12 Lyn AB	5.4, 6.0	1.7	69	Σ948	AB is binary, P = 706 years.
					12 Lyn AC	5.4, 7.3	8.7	309		

No.	RA		Declin-ation		Star	Magni-tudes	Separa-tion	PA	Cata-logue	Comments
	h	m	°	′			arcsec	°		
30	07	08.7	−70	30	γ Vol	3.9, 5.8	14.1	298	Δ42	Very slow binary.
31	07	16.6	−23	19	h3945 CMa	4.8, 6.8	26.8	51	−	Contrasting colours.
32	07	20.1	+21	59	δ Gem	3.5, 8.2	5.7	226	Σ1066	Not too easy. Yellow, pale blue.
33	07	34.6	+31	53	α Gem	1.9, 2.9	4.2	61	Σ1110	Widening. Easy with 7.5 cm.
34	07	38.8	−26	48	κ Pup	4.5, 4.7	9.8	318	H III 27	Both white.
35	08	12.2	+17	39	ζ Cnc AB	5.6, 6.0	0.9	62	Σ1196	Period (AB) = 60 years. Near maximum separation.
					ζ Cnc AB-C	5.0, 6.2	5.9	71	Σ1196	Period (AB-C) = 1,150 years.
36	08	44.7	−54	43	δ Vel	2.1, 5.1	0.9	334	I 10	Difficult close pair. Period 142 years.
37	08	46.8	+06	25	ε Hyd	3.3, 6.8	2.9	301	Σ1273	PA slowly increasing. A is a very close pair.
38	09	18.8	+36	48	38 Lyn	3.9, 6.6	2.8	230	Σ1338	Almost fixed.
39	09	47.1	−65	04	μ Car	3.1, 6.1	5.0	128	RMK 11	Fixed. Fine in small telescopes.
40	10	20.0	+19	50	γ Leo	2.2, 3.5	4.4	125	Σ1424	Binary, period = 619 years. Both orange.
41	10	32.0	−45	04	s Vel	6.2, 6.5	13.5	218	PZ 3	Fixed.
42	10	46.8	−49	26	μ Vel	2.7, 6.4	2.5	54	R 155	P = 138 years. Near widest separation.
43	10	55.6	+24	45	54 Leo	4.5, 6.3	6.6	111	Σ1487	Slowly widening. Pale yellow and white.

No.	RA		Declin-ation		Star	Magni-tudes	Separa-tion	PA	Cata-logue	Comments
	h	m	°	′			arcsec	°		
44	11	18.2	+31	32	ξ UMa	4.3, 4.8	1.8	252	Σ1523	Binary, 60 years. Needs 7.5 cm.
45	11	21.0	−54	29	π Cen	4.3, 5.0	0.3	152	I 879	Binary, 38.7 years. Very close. Needs 35 cm.
46	11	23.9	+10	32	ι Leo	4.0, 6.7	1.8	107	Σ1536	Binary, period = 186 years. Slowly widening.
47	11	32.3	−29	16	N Hya	5.8, 5.9	9.5	210	H III 96	Fixed.
48	12	14.0	−45	43	D Cen	5.6, 6.8	2.8	243	RMK 14	Orange and white. Closing.
49	12	26.6	−63	06	α Cru	1.4, 1.9	4.0	112	Δ252	Third star in a low-power field.
50	12	41.5	−48	58	γ Cen	2.9, 2.9	0.8	343	HJ 4539	Period = 84 years. Closing. Both yellow.
51	12	41.7	−01	27	γ Vir	3.5, 3.5	0.6	216	Σ1670	Binary. P = 169 years. Reaches 0″.5 in early 2005.
52	12	46.3	−68	06	β Mus	3.7, 4.0	1.3	45	R 207	Both white. Closing slowly. P = 383 years.
53	12	54.6	−57	11	μ Cru	4.3, 5.3	34.9	17	Δ126	Fixed. Both white.
54	12	56.0	+38	19	α CVn	2.9, 5.5	19.3	229	Σ1692	Easy. Yellow, bluish.
55	13	22.6	−60	59	J Cen	4.6, 6.5	60.0	343	Δ133	Fixed. A is a close pair.
56	13	24.0	+54	56	ζ UMa	2.3, 4.0	14.4	152	Σ1744	Very easy. Naked-eye pair with Alcor.
57	13	51.8	−33	00	3 Cen	4.5, 6.0	7.9	106	H III 101	Both white. Closing slowly.

No.	RA		Declin-ation		Star	Magni-tudes	Separa-tion	PA	Cata-logue	Comments
	h	m	°	′			arcsec	°		
58	14	39.6	−60	50	α Cen	0.0, 1.2	11.3	228	RHD 1	Finest pair in the sky. P = 80 years. Closing.
59	14	41.1	+13	44	ζ Boo	4.5, 4.6	0.7	298	Σ1865	Both white. Closing – highly inclined orbit.
60	14	45.0	+27	04	ε Boo	2.5, 4.9	2.9	345	Σ1877	Yellow, blue. Fine pair.
61	14	46.0	−25	27	54 Hya	5.1, 7.1	8.3	122	H III 97	Closing slowly.
62	14	49.3	−14	09	μ Lib	5.8, 6.7	1.9	2	β106	Becoming wider. Fine in 7.5 cm.
63	14	51.4	+19	06	ξ Boo	4.7, 7.0	6.4	314	Σ1888	Fine contrast. Easy.
64	15	03.8	+47	39	44 Boo	5.3, 6.2	2.2	55	Σ1909	Period = 246 years.
65	15	05.1	−47	03	π Lup	4.6, 4.7	1.7	67	HJ 4728	Widening.
66	15	18.5	−47	53	μ Lup AB	5.1, 5.2	0.8	120	HJ 4753	AB closing. Underobserved.
					μ Lup AC	4.4, 7.2	24.0	129	Δ180	AC almost fixed.
67	15	23.4	−59	19	γ Cir	5.1, 5.5	0.8	353	HJ 4757	Closing. Needs 20 cm. Long-period binary.
68	15	32.0	+32	17	η CrB	5.6, 5.9	0.5	98	Σ1937	Both yellow. P = 41 yrs. Closing.
69	15	34.8	+10	33	δ Ser	4.2, 5.2	4.3	176	Σ1954	Long-period binary.
70	15	35.1	−41	10	γ Lup	3.5, 3.6	0.8	278	HJ 4786	Binary. Period = 190 years. Needs 20 cm.
71	15	56.9	−33	58	ξ Lup	5.3, 5.8	10.2	49	PZ 4	Fixed.
72	16	14.7	+33	52	σ CrB	5.6, 6.6	7.1	237	Σ2032	Long-period binary. Both white.
73	16	29.4	−26	26	α Sco	1.2, 5.4	2.6	274	GNT 1	Red, green. Difficult from mid-northern latitudes.

No.	RA		Declin-ation		Star	Magni-tudes	Separa-tion	PA	Cata-logue	Comments
	h	m	°	′			arcsec	°		
74	16	30.9	+01	59	λ Oph	4.2, 5.2	1.5	32	Σ2055	P = 129 years. Fairly difficult in small apertures.
75	16	41.3	+31	36	ζ Her	2.9, 5.5	0.8	240	Σ2084	Period 34 years. Now widening. Needs 25 cm.
76	17	05.3	+54	28	μ Dra	5.7, 5.7	2.3	13	Σ2130	Period 672 years.
77	17	14.6	+14	24	α Her	var, 5.4	4.6	104	Σ2140	Red, green. Long-period binary.
78	17	15.3	−26	35	36 Oph	5.1, 5.1	4.9	326	SHJ 243	Period = 471 years.
79	17	23.7	+37	08	ρ Her	4.6, 5.6	4.1	318	Σ2161	Slowly widening.
80	18	01.5	+21	36	95 Her	5.0, 5.1	6.4	257	Σ2264	Colours thought variable in C19.
81	18	05.5	+02	30	70 Oph	4.2, 6.0	4.7	140	Σ2272	Opening. Easy in 7.5 cm.
82	18	06.8	−43	25	h5014 CrA	5.7, 5.7	1.7	4	–	Period = 450 years. Needs 10 cm.
83	18	35.9	+16	58	OΣ358 Her	6.8, 7.0	1.6	152	–	Period = 380 years.
84	18	44.3	+39	40	ε¹ Lyr	5.0, 6.1	2.5	349	Σ2382	Quadruple system with ε². Both pairs
85	18	44.3	+39	40	ε² Lyr	5.2, 5.5	2.4	81	Σ2383	visible in 7.5 cm.
86	18	56.2	+04	12	θ Ser	4.5, 5.4	22.4	104	Σ2417	Fixed. Very easy.
87	19	06.4	−37	04	γ CrA	4.8, 5.1	1.3	42	HJ 5084	Beautiful pair. Period = 122 years. Slowly widening.
88	19	30.7	+27	58	β Cyg AB	3.1, 5.1	34.3	54	Σ I 43	Glorious. Yellow, blue-greenish.
					β Cyg Aa	3.1, 4.0	0.3	115	MCA 55	Aa. Discovered in 1976. Period = 97 years.

No.	RA		Declin-ation		Star	Magni-tudes	Separa-tion	PA	Cata-logue	Comments
	h	m	°	′			arcsec	°		
89	19	45.0	+45	08	δ Cyg	2.9, 6.3	2.6	223	Σ2579	Slowly widening. Period = 780 years.
90	19	48.2	+70	16	ε Dra	3.8, 7.4	3.2	17	Σ2603	Slow binary.
91	20	46.7	+16	07	γ Del	4.5, 5.5	9.2	266	Σ2727	Easy. Yellowish. Long-period binary.
92	20	47.4	+36	29	λ Cyg	4.8, 6.1	0.9	10	OΣ413	Difficult binary in small apertures.
93	20	59.1	+04	18	ε Equ AB	6.0, 6.3	0.7	284	Σ2737	Fine triple. AB is closing.
					ε Equ AC	6.0, 7.1	10.3	66		
94	21	06.9	+38	45	61 Cyg	5.2, 6.0	30.8	150	Σ2758	Nearby binary. Both orange. Period = 722 years.
95	21	19.9	−53	27	θ Ind	4.5, 7.0	6.8	271	HJ 5258	Pale yellow and reddish. Long-period binary.
96	21	44.1	+28	45	μ Cyg	4.8, 6.1	1.8	311	Σ2822	Period = 713 years.
97	22	03.8	+64	37	ξ Cep	4.4, 6.5	8.1	275	Σ2863	White and blue. Long-period binary.
98	22	26.6	−16	45	53 Aqr	6.4, 6.6	1.5	17	SHJ 345	Long-period binary, approaching periastron.
99	22	28.8	−00	01	ζ Aqr	4.3, 4.5	2.0	179	Σ2909	Slowly widening.
100	23	59.4	+33	43	Σ3050 And	6.6, 6.6	2.0	331	–	Period = 350 years.

Some Interesting Nebulae, Clusters and Galaxies

Object	RA		Declina-tion		Remarks
	h	m	°	′	
M31 Andromedae	00	40.7	+41	05	Andromeda Galaxy, visible to naked eye.
H VIII 78 Cassiopeiae	00	41.3	+61	36	Fine cluster, between Gamma and Kappa Cassiopeiae.
M33 Trianguli	01	31.8	+30	28	Spiral. Difficult with small apertures.
H VI 33–4 Persei, C14	02	18.3	+56	59	Double cluster; Sword-handle.
Δ142 Doradus	05	39.1	−69	09	Looped nebula round 30 Doradus. Naked eye. In Large Magellanic Cloud.
M1 Tauri	05	32.3	+22	00	Crab Nebula, near Zeta Tauri.
M42 Orionis	05	33.4	−05	24	Orion Nebula. Contains the famous Trapezium, Theta Orionis.
M35 Geminorum	06	06.5	+24	21	Open cluster near Eta Geminorum.
H VII 2 Monocerotis, C50	06	30.7	+04	53	Open cluster, just visible to naked eye.
M41 Canis Majoris	06	45.5	−20	42	Open cluster, just visible to naked eye.
M47 Puppis	07	34.3	−14	22	Mag. 5.2. Loose cluster.
H IV 64 Puppis	07	39.6	−18	05	Bright planetary in rich neighbourhood.
M46 Puppis	07	39.5	−14	42	Open cluster.
M44 Cancri	08	38	+20	07	Praesepe. Open cluster near Delta Cancri. Visible to naked eye.
M97 Ursae Majoris	11	12.6	+55	13	Owl Nebula, diameter 3′. Planetary.
Kappa Crucis, C94	12	50.7	−60	05	'Jewel Box'; open cluster, with stars of contrasting colours.
M3 Can. Ven.	13	40.6	+28	34	Bright globular.
Omega Centauri, C80	13	23.7	−47	03	Finest of all globulars. Easy with naked eye.
M80 Scorpii	16	14.9	−22	53	Globular, between Antares and Beta Scorpii.
M4 Scorpii	16	21.5	−26	26	Open cluster close to Antares.
M13 Herculis	16	40	+36	31	Globular. Just visible to naked eye.

Object	RA		Declina-tion		Remarks
	h	m	°	′	
M92 Herculis	16	16.1	+43	11	Globular. Between Iota and Eta Herculis.
M6 Scorpii	17	36.8	−32	11	Open cluster; naked eye.
M7 Scorpii	17	50.6	−34	48	Very bright open cluster; naked eye.
M23 Sagittarii	17	54.8	−19	01	Open cluster nearly 50′ in diameter.
H IV 37 Draconis, C6	17	58.6	+66	38	Bright planetary.
M8 Sagittarii	18	01.4	−24	23	Lagoon Nebula. Gaseous. Just visible with naked eye.
NGC 6572 Ophiuchi	18	10.9	+06	50	Bright planetary, between Beta Ophiuchi and Zeta Aquilae.
M17 Sagittarii	18	18.8	−16	12	Omega Nebula. Gaseous. Large and bright.
M11 Scuti	18	49.0	−06	19	Wild Duck. Bright open cluster.
M57 Lyrae	18	52.6	+32	59	Ring Nebula. Brightest of planetaries.
M27 Vulpeculae	19	58.1	+22	37	Dumb-bell Nebula, near Gamma Sagittae.
H IV 1 Aquarii, C55	21	02.1	−11	31	Bright planetary, near Nu Aquarii.
M15 Pegasi	21	28.3	+12	01	Bright globular, near Epsilon Pegasi.
M39 Cygni	21	31.0	+48	17	Open cluster between Deneb and Alpha Lacertae. Well seen with low powers.

(M = Messier number; NGC = New General Catalogue number; C = Caldwell number.)

Our Contributors

Dr Paul Murdin was formerly Head of Astronomy at the Particle Physics and Astronomy Research Council (PPARC) and Director of Science at the British National Space Centre. He now works at the Institute of Astronomy in Cambridge. He is, of course, one of our most regular contributors.

David M. Harland gained his BSc in astronomy in 1977 and a doctorate in computational science. Subsequently, he has taught computer science, worked in industry and managed academic research. In 1995 he 'retired' and has since published many books on space themes, including '*Mission To Saturn – Cassini and the Huygens Probe*' in 2002.

Professor Chris Kitchin was formerly Director of the University of Hertfordshire Observatory. He is an astrophysicist with a great eagerness in encouraging a popular interest in astronomy. He is the author of several books, and appears regularly on television.

Dr S. J. Wainwright is a senior lecturer in Environmental Biology at the University of Wales, Swansea. He is a fellow of the Royal Astronomical Society and a member of the Cardiff and Swansea Astronomical Societies. He is the founder member of the QCUIAG Imaging group.

Michael T. Wright, curator of Mechanical Engineering at the Science Museum, London, has worked there for thirty-two years. He spends his leisure time making and mending delicate mechanisms such as clocks, and musical and scientific instruments. His studies in ancient Greek technology combine his enthusiasm for practical work with interests in mathematics and the classics.

Fred Taylor is Halley Professor of Physics at Oxford University and was for many years the Head of the Atmospheric, Oceanic and Planetary Physics subdepartment. He is also a Distinguished Visiting Scientist at the Jet Propulsion Laboratory of the California Institute of Technology in the USA, and President of the International Commission for Planetary Atmospheres and their Evolution in the IUGG (International Union of Geodesy and Geophysics). His research interests centre around space missions to the planets, including Earth satellites, and experiments for the investigation of planetary atmospheres and climate.

David A. Hardy is European Vice-President of the IAAA and recipient in 2001 of the Rudaux Memorial Award for services to astronomical art. He is author/compiler of *Visions of Space: Artists Journey Through the Cosmos* (Dragon's World, 1989/1990). A book about his life and work, *Hardyware: The Art of David A. Hardy*, was published by Paper Tiger (Collins & Brown) in September 2001. In March 2003, an asteroid was named after him: (13329) Davidhardy = 1998 SB32.

Professor Fred Watson is Astronomer-in-Charge of the Anglo-Australian Observatory at Coonabarabran, New South Wales. He is an Adjunct Professor in the School of Physical and Chemical Sciences of the Queensland University of Technology, and an honorary Associate Professor of Astronomy in the University of Southern Queensland.

Dr Allan Chapman of Wadham College, Oxford, is probably Britain's leading authority on the history of astronomy. He has published many research papers and several books, as well as numerous popular accounts. He is a frequent and welcome contributor to the *Yearbook*.

Astronomical Societies in the British Isles

British Astronomical Association
Assistant Secretary: Burlington House, Piccadilly, London W1V 9AG.
Meetings: Lecture Hall of Scientific Societies, Civil Service Commission Building, 23 Savile Row, London W1. Last Wednesday each month (Oct.–June), 5 p.m. and some Saturday afternoons.

Association for Astronomy Education
Secretary: Teresa Grafton, The Association for Astronomy Education, c/o The Royal Astronomical Society, Burlington House, Piccadilly, London W1V 0NL.

Astronomical Society of Edinburgh
Secretary: Graham Rule, 105/19 Causewayside, Edinburgh EH9 1QG.
Web site: www.roe.ac.uk/asewww/; *Email:* asewww@roe.ac.uk
Meetings: City Observatory, Calton Hill, Edinburgh. 1st Friday each month, 8 p.m.

Astronomical Society of Glasgow
Secretary: Mr Robert Hughes, Apartment 8/4, 75 Plean Street, Glasgow G14 0YW.
Meetings: University of Strathclyde, George St, Glasgow. 3rd Thursday each month, Sept.–Apr., 7.30 p.m.

Astronomical Society of Haringey
Secretary: Jerry Workman, 91 Greenslade Road, Barking, Essex, IG11 9XF.
Meetings: Palm Court, Alexandra Palace, 3rd Wednesday each month, 8 p.m.

Astronomy Ireland
Secretary: Tony Ryan, PO Box 2888, Dublin 1, Eire.
Web site: www.astronomy.ie; *Email:* info@astronomy.ie
Meetings: 2nd Monday of each month. Telescope meetings every clear Saturday.

Federation of Astronomical Societies
Secretary: Clive Down, 10 Glan-y-Llyn, North Cornelly, Bridgend, County Borough, CF33 4EF.
Email: clivedown@btinternet.com

Junior Astronomical Society of Ireland
Secretary: K. Nolan, 5 St Patrick's Crescent, Rathcoole, Co. Dublin.
Meetings: The Royal Dublin Society, Ballsbridge, Dublin 4. Monthly.

Society for Popular Astronomy
Secretary: Guy Fennimore, 36 Fairway, Keyworth, Nottingham, NG12 5DU.
Web site: www.popastro.com; *Email:* SPAstronomy@aol.com
Meetings: Last Saturday in Jan., Apr., July, Oct., 2.30 p.m. in London.

Webb Society
Secretary: M. B. Swan, Carrowreagh, Kilshanny, Kilfenora, Co. Clare, Eire.

Aberdeen and District Astronomical Society
Secretary: Ian C. Giddings, 95 Brentfield Circle, Ellon, Aberdeenshire AB41 9DB.
Meetings: Robert Gordon's Institute of Technology, St Andrew's Street, Aberdeen.
Fridays, 7.30 p.m.

Abingdon Astronomical Society (was **Fitzharry's Astronomical Society**)
Secretary: Chris Holt, 9 Rutherford Close, Abingdon, Oxon OX14 2AT.
Web site: www.abingdonastro.org.uk; *Email:* info@abingdonastro.co.uk
Meetings: All Saints' Methodist Church Hall, Dorchester Crescent, Abingdon, Oxon.
2nd Monday Sept.–June, 8 p.m. and additional beginners' meetings and observing
evenings as advertised.

Altrincham and District Astronomical Society
Secretary: Derek McComiskey, 33 Tottenham Drive, Manchester M23 9WH.
Meetings: Timperley Village Club. 1st Friday Sept.–June, 8 p.m.

Andover Astronomical Society
Secretary: Mrs S. Fisher, Staddlestones, Aughton, Kingston, Marlborough, Wiltshire,
SN8 3SA.
Meetings: Grately Village Hall. 3rd Thursday each month, 7.30 p.m.

Astra Astronomy Section
Secretary: c/o Duncan Lunan, Flat 65, Dalraida House, 56 Blythswood Court,
Anderston, Glasgow G2 7PE.
Meetings: Airdrie Arts Centre, Anderson Street, Airdrie. Weekly.

Astrodome Mobile School Planetarium
Contact: Peter J. Golding, 53 City Way, Rochester, Kent ME1 2AX.
Web site: www.astrodome.clara.co.uk; *Email:* astrodome@clara.co.uk

Aylesbury Astronomical Society
Secretary: Alan Smith, 182 Marley Fields, Leighton Buzzard, Beds, LU7 8WN.
Meetings: 1st Monday in month at 8 p.m., venue in Aylesbury area. Details from
Secretary.

Bassetlaw Astronomical Society
Secretary: Andrew Patton, 58 Holding, Worksop, Notts S81 0TD.
Meetings: Rhodesia Village Hall, Rhodesia, Worksop, Notts. 2nd and 4th Tuesdays of
month at 7.45 p.m.

Batley & Spenborough Astronomical Society
Secretary: Robert Morton, 22 Links Avenue, Cleckheaton, West Yorks BD19 4EG.
Meetings: Milner K. Ford Observatory, Wilton Park, Batley. Every Thursday, 8 p.m.

Bedford Astronomical Society
Secretary: Mrs L. Harrington, 24 Swallowfield, Wyboston, Bedfordshire, MK44 3AE.
Web site: www.observer1.freeserve.co.uk/bashome.html
Meetings: Bedford School, Burnaby Rd, Bedford. Last Wednesday each month.

Bingham & Brooks Space Organization
Secretary: N. Bingham, 15 Hickmore's Lane, Lindfield, W. Sussex.

Birmingham Astronomical Society
Contact: P. Bolas, 4 Moat Bank, Bretby, Burton on Trent DE15 0QJ.
Web site: www.birmingham-astronomical.co.uk; *Email:* pbolas@aol.com
Meetings: Room 146, Aston University. Last Tuesday of month. Sept.–June (except
Dec., moved to 1st week in Jan.).

Blackburn Leisure Astronomy Section
Secretary: Mr H. Murphy, 20 Princess Way, Beverley, East Yorkshire, HU17 8PD.
Meetings: Blackburn Leisure Welfare. Mondays, 8 p.m.

Blackpool & District Astronomical Society
Secretary: Terry Devon, 30 Victory Road, Blackpool, Lancashire, FY1 3JT.
Acting Secretary: Tony Evanson, 25 Aintree Road, Thornton, Lancashire, FY5 5HW.
Web site: www.geocities.com/bad_astro/index.html; *Email:* bad_astro@yahoo.co.uk
Meetings: St Kentigens Social Centre, Blackpool. 1st Wednesday of the month, 8 p.m.

Bolton Astronomical Society
Secretary: Peter Miskiw, 9 Hedley Street, Bolton, Lancashire, BL1 3LE.
Meetings: Ladybridge Community Centre, Bolton. 1st and 3rd Tuesdays Sept.–May,
7.30 p.m.

Border Astronomy Society
Secretary: David Pettitt, 14 Sharp Grove, Carlisle, Cumbria, CA2 5QR.
Web site: www.members.aol.com/P3pub/page8.html
Email: davidpettitt@supanet.com
Meetings: The Observatory, Trinity School, Carlisle. Alternate Thursdays, 7.30 p.m.,
Sept.–May.

Boston Astronomers
Secretary: Mrs Lorraine Money, 18 College Park, Horncastle, Lincolnshire, LN9 6RE.
Meetings: Blackfriars Arts Centre, Boston. 2nd Monday each month, 7.30 p.m.

Bradford Astronomical Society
Contact: Mrs J. Hilary Knaggs, 6 Meadow View, Wyke, Bradford, BD12 9LA.
Web site: www.bradford-astro.freeserve.co.uk/index.htm
Meetings: Eccleshill Library, Bradford. Alternate Mondays, 7.30 p.m.

Braintree, Halstead & District Astronomical Society
Secretary: Mr J. R. Green, 70 Dorothy Sayers Drive, Witham, Essex, CM8 2LU.
Meetings: BT Social Club Hall, Witham Telephone Exchange. 3rd Thursday each
month, 8 p.m.

Breckland Astronomical Society (was **Great Ellingham and District Astronomy Club**)
Contact: Martin Wolton, Willowbeck House, Pulham St Mary, Norfolk, IP21 4QS.
Meetings: Great Ellingham Recreation Centre, Watton Road (B1077), Great
Ellingham, 2nd Friday each month, 7.15 p.m.

Bridgend Astronomical Society
Secretary: Clive Down, 10 Glan-y-Llyn, Broadlands, North Cornelly, Bridgend
County, CF33 4EF.
Email: clivedown@btinternet.com
Meetings: Bridgend Bowls Centre, Bridgend. 2nd Friday, monthly, 7.30 p.m.

Bridgwater Astronomical Society
Secretary: Mr G. MacKenzie, Watergore Cottage, Watergore, South Petherton,
Somerset, TA13 5JQ.
Web site: www.ourworld.compuserve.com/hompages/dbown/Bwastro.htm
Meetings: Room D10, Bridgwater College, Bath Road Centre, Bridgwater. 2nd
Wednesday each month, Sept.–June.

Bridport Astronomical Society
Secretary: Mr G. J. Lodder, 3 The Green, Walditch, Bridport, Dorset, DT6 4LB.
Meetings: Walditch Village Hall, Bridport. 1st Sunday each month, 7.30 p.m.

Brighton Astronomical and Scientific Society
Secretary: Ms T. Fearn, 38 Woodlands Close, Peacehaven, East Sussex, BN10 7SF.
Meetings: St Johns Church Hall, Hove. 1st Tuesday each month, 7.30 p.m.

Bristol Astronomical Society
Secretary: Dr John Pickard, 'Fielding', Easter Compton, Bristol, BS35 5SJ.
Meetings: Frank Lecture Theatre, University of Bristol Physics Dept., alternate
Fridays in term time, and Westbury Park Methodist Church Rooms, North View,
other Fridays.

Cambridge Astronomical Society
Secretary: Brian Lister, 80 Ramsden Square, Cambridge CB4 2BL.
Meetings: Institute of Astronomy, Madingley Road. 3rd Friday each month.

Cardiff Astronomical Society
Secretary: D. W. S. Powell, 1 Tal-y-Bont Road, Ely, Cardiff CF5 5EU.
Meetings: Dept. of Physics and Astronomy, University of Wales, Newport Road,
Cardiff. Alternate Thursdays, 8 p.m.

Castle Point Astronomy Club
Secretary: Andrew Turner, 3 Canewdon Hall Close, Canewdon, Rochford, Essex
SS4 3PY.
Meetings: St Michael's Church Hall, Daws Heath. Wednesdays, 8 p.m.

Chelmsford Astronomers
Secretary: Brendan Clark, 5 Borda Close, Chelmsford, Essex.
Meetings: Once a month.

Chester Astronomical Society
Secretary: Mrs S. Brooks, 39 Halton Road, Great Sutton, South Wirral, LL66 2UF.
Meetings: All Saints Parish Church, Chester. Last Wednesday each month except
Aug. and Dec., 7.30 p.m.

Chester Society of Natural Science, Literature and Art
Secretary: Paul Braid, 'White Wing', 38 Bryn Avenue, Old Colwyn, Colwyn Bay
LL29 8AH.
Email: p.braid@virgin.net
Meetings: Once a month.

Chesterfield Astronomical Society
President: Mr D. Blackburn, 71 Middlecroft Road, Stavely, Chesterfield, Derbyshire,
S41 3XG. Tel: 07909 570754.
Website: www.chesterfield-as.org.uk
Meetings: Barnet Observatory, Newbold, each Friday.

Clacton & District Astronomical Society
Secretary: C. L. Haskell, 105 London Road, Clacton-on-Sea, Essex.

Cleethorpes & District Astronomical Society
Secretary: C. Illingworth, 38 Shaw Drive, Grimsby, S. Humberside.
Meetings: Beacon Hill Observatory, Cleethorpes. 1st Wednesday each month.

Cleveland & Darlington Astronomical Society
Contact: Dr. John McCue, 40 Bradbury Rd., Stockton-on-Tees, Cleveland TS20 1LE.
Meetings: Grindon Parish Hall, Thorpe Thewles, near Stockton-on-Tees. 2nd Friday,
monthly.

Cork Astronomy Club
Secretary: Charles Coughlan, 12 Forest Ridge Crescent, Wilton, Cork, Eire.
Meetings: 1st Monday, Sept.–May (except bank holidays).

Cornwall Astronomical Society
Secretary: J. M. Harvey, 1 Tregunna Close, Porthleven, Cornwall TR13 9LW.
Meetings: Godolphin Club, Wendron Street, Helston, Cornwall. 2nd and 4th
Thursday of each month, 7.30 for 8 p.m.

Cotswold Astronomical Society

Secretary: Rod Salisbury, Grove House, Christchurch Road, Cheltenham, Glos GL50 2PN.

Web site: www.members.nbci.com/CotswoldAS

Meetings: Shurdington Church Hall, School Lane, Shurdington, Cheltenham. 2nd Saturday each month, 8 p.m.

Coventry & Warwickshire Astronomical Society

Secretary: Steve Payne, 68 Stonebury Avenue, Eastern Green, Coventry CV5 7FW.

Web site: www.cawas.freeserve.co.uk; *Email:* sjp2000@thefarside57.freeserve.co.uk

Meetings: The Earlsdon Church Hall, Albany Road, Earlsdon, Coventry. 2nd Friday, monthly, Sept.–June.

Crawley Astronomical Society

Secretary: Ron Gamer, 1 Pevensey Close, Pound Hill, Crawley, West Sussex RH10 7BL.

Meetings: Ifield Community Centre, Ifield Road, Crawley. 3rd Friday each month, 7.30 p.m.

Crayford Manor House Astronomical Society

Secretary: Roger Pickard, 28 Appletons, Hadlow, Kent TM1 0DT.

Meetings: Manor House Centre, Crayford. Monthly during term time.

Croydon Astronomical Society

Secretary: John Murrell, 17 Dalmeny Road, Carshalton, Surrey.

Meetings: Lecture Theatre, Royal Russell School, Combe Lane, South Croydon. Alternate Fridays, 7.45 p.m.

Derby & District Astronomical Society

Secretary: Ian Bennett, Freers Cottage, Sutton Lane, Etwall.

Web site: www.derby-astro-soc.fsnet/index.html

Email: bennett.lovatt@btinternet.com

Meetings: Friends Meeting House, Derby. 1st Friday each month, 7.30 p.m.

Doncaster Astronomical Society

Secretary: A. Anson, 15 Cusworth House, St James Street, Doncaster, DN1 3AY

Web site: www.donastro.freeserve.co.uk

Email: space@donastro.freeserve.co.uk

Meetings: St George's Church House, St George's Church, Church Way, Doncaster. 2nd and 4th Thursday of each month, commencing at 7.30 p.m.

Dumfries Astronomical Society

Secretary: Mr J. Sweeney, 3 Lakeview, Powfoot, Annan, DG13 5PG.

Meetings: Gracefield Arts Centre, Edinburgh Road, Dumfries. 3rd Tuesday Aug.–May, 7.30 p.m.

Dundee Astronomical Society

Secretary: G. Young, 37 Polepark Road, Dundee, Tayside, DD1 5QT.

Meetings: Mills Observatory, Balgay Park, Dundee. 1st Friday each month, 7.30 p.m. Sept.–Apr.

Easington and District Astronomical Society

Secretary: T. Bradley, 52 Jameson Road, Hartlepool, Co. Durham.

Meetings: Easington Comprehensive School, Easington Colliery. Every 3rd Thursday throughout the year, 7.30 p.m.

Eastbourne Astronomical Society
Secretary: Peter Gill, 18 Selwyn House, Selwyn Road, Eastbourne, East Sussex
BN21 2LF.
Meetings: Willingdon Memorial Hall, Church Street, Willingdon. One Saturday per
month, Sept.–July, 7.30 p.m.

East Riding Astronomers
Secretary: Tony Scaife, 15 Beech Road, Elloughton, Brough, North Humberside,
HU15 1JX.
Meetings: As arranged.

East Sussex Astronomical Society
Secretary: Marcus Croft, 12 St Marys Cottages, Ninfield Road, Bexhill on Sea, East
Sussex.
Web site: www.esas.org.uk
Meetings: St Marys School, Wrestwood Road, Bexhill. 1st Thursday of each month,
8 p.m.

Edinburgh University Astronomical Society
Secretary: c/o Dept. of Astronomy, Royal Observatory, Blackford Hill, Edinburgh.

Ewell Astronomical Society
Secretary: Richard Gledhill, 80 Abinger Avenue, Cheam SM2 7LW.
Web site: www.ewell-as.co.uk
Meetings: St Mary's Church Hall, London Road, Ewell. 2nd Friday of each month
except August, 7.45 p.m.

Exeter Astronomical Society
Secretary: Tim Sedgwick, Old Dower House, Half Moon, Newton St Cyres, Exeter,
Devon, EX5 5AE.
Meetings: The Meeting Room, Wynards, Magdalen Street, Exeter. 1st Thursday of
month.

Farnham Astronomical Society
Secretary: Laurence Anslow, 'Asterion', 18 Wellington Lane, Farnham, Surrey,
GU9 9BA.
Meetings: Central Club, South Street, Farnham. 2nd Thursday each month, 8 p.m.

Foredown Tower Astronomy Group
Secretary: M. Feist, Foredown Tower Camera Obscura, Foredown Road, Portslade,
East Sussex BN41 2EW.
Meetings: At the above address, 3rd Tuesday each month. 7 p.m. (winter), 8 p.m.
(summer).

Fylde Astronomical Society
Secretary: 28 Belvedere Road, Thornton, Lancs.
Meetings: Stanley Hall, Rossendale Avenue South. 1st Wednesday each month.

Greenock Astronomical Society
Secretary: Carl Hempsey, 49 Brisbane Street, Greenock.
Meetings: Greenock Arts Guild, 3 Campbell Street, Greenock.

Grimsby Astronomical Society
Secretary: R. Williams, 14 Richmond Close, Grimsby, South Humberside.
Meetings: Secretary's home. 2nd Thursday each month, 7.30 p.m.

Guernsey: La Société Guernesiasie Astronomy Section
Secretary: Debby Quertier, Lamorna, Route Charles, St Peter Port, Guernsey GY1
1QS and Jessica Harris, Keanda, Les Sauvagees, St Sampsons, Guernsey GY2 4XT.
Meetings: Observatory, Rue du Lorier, St Peters. Tuesdays, 8 p.m.

Guildford Astronomical Society
Secretary: A. Langmaid, 22 West Mount, The Mount, Guildford, Surrey, GU2 5HL.
Meetings: Guildford Institute, Ward Street, Guildford. 1st Thursday each month, except Aug., 7.30 p.m.

Gwynedd Astronomical Society
Secretary: Mr Ernie Greenwood, 18 Twrcelyn Street, Llanerchymedd, Anglesey LL74 8TL.
Meetings: Dept. of Electronic Engineering, Bangor University. 1st Thursday each month except Aug., 7.30 p.m.

The Hampshire Astronomical Group
Secretary: Geoff Mann, 10 Marie Court, 348 London Road, Waterlooville, Hants PO7 7SR.
Web site: www.hantsastro.demon.co.uk; *Email:* Geoff.Mann@hazleton97.fsnet.co.uk
Meetings: 2nd Friday, Clanfield Memorial Hall, all other Fridays Clanfield Observatory.

Hanney & District Astronomical Society
Secretary: Bob Church, 47 Upthorpe Drive, Wantage, Oxfordshire, OX12 7DG.
Meetings: Last Thursday each month, 8 p.m.

Harrogate Astronomical Society
Secretary: Brian Bonser, 114 Main Street, Little Ouseburn, TO5 9TG.
Meetings: National Power HQ, Beckwith Knowle, Harrogate. Last Friday each month.

Hastings and Battle Astronomical Society
Secretary: K. A. Woodcock, 24 Emmanuel Road, Hastings, East Sussex, TN34 3LB.
Email: keith@habas.freeserve.co.uk
Meetings: Herstmonceux Science Centre. 2nd Saturday of each month, 7.30 p.m.

Havering Astronomical Society
Secretary: Frances Ridgley, 133 Severn Drive, Upminster, Essex, RM14 1PP.
Meetings: Cranham Community Centre, Marlborough Gardens, Upminster, Essex. 3rd Wednesday each month (except July and Aug.), 7.30 p.m.

Heart of England Astronomical Society
Secretary: John Williams, 100 Stanway Road, Shirley, Solihull, B90 3JG.
Web site: www.members.aol.com/hoeas/home.html; *Email:* hoeas@aol.com
Meetings: Furnace End Village, over Whitacre, Warwickshire. Last Thursday each month, except June, July & Aug., 8 p.m.

Hebden Bridge Literary & Scientific Society, Astronomical Section
Secretary: Peter Jackson, 44 Gilstead Lane, Bingley, West Yorkshire, BD16 3NP.
Meetings: Hebden Bridge Information Centre. Last Wednesday, Sept.–May.

Herschel Astronomy Society
Secretary: Kevin Bishop, 106 Holmsdale, Crown Wood, Bracknell, Berkshire, RG12 3TB.
Meetings: Eton College. 2nd Friday each month, 7.30 p.m.

Highlands Astronomical Society
Secretary: Richard Green, 11 Drumossie Avenue, Culcabock, Inverness IV2 3SJ.
Meetings: The Spectrum Centre, Inverness. 1st Tuesday each month, 7.30 p.m.

Hinckley & District Astronomical Society
Secretary: Mr S. Albrighton, 4 Walnut Close, The Bridleways, Hartshill, Nuneaton, Warwickshire, CV10 0XH.
Meetings: Burbage Common Visitors Centre, Hinckley. 1st Tuesday Sept.–May, 7.30 p.m.

Horsham Astronomy Group (was **Forest Astronomical Society**)
Secretary: Mr A. R. Clarke, 93 Clarence Road, Horsham, West Sussex, RH13 5SL.
Meetings: 1st Wednesday each month.

Howards Astronomy Club
Secretary: H. Ilett, 22 St Georges Avenue, Warblington, Havant, Hants.
Meetings: To be notified.

Huddersfield Astronomical and Philosophical Society
Secretary: Lisa B. Jeffries, 58 Beaumont Street, Netherton, Huddersfield, West Yorkshire, HD4 7HE.
Email: l.b.jeffries@hud.ac.uk
Meetings: 4a Railway Street, Huddersfield. Every Wednesday and Friday, 7.30 p.m.

Hull and East Riding Astronomical Society
Secretary: Tony Scaife, 15 Beech Road, Elloughton, Brough, North Humberside, HU15 1JX.
Meetings: Wyke 6th Form College, Bricknell Avenue, Hull. 2nd Tuesday each month, Oct.–Apr., 7.30 p.m.

Ilkeston & District Astronomical Society
Secretary: Mark Thomas, 2 Elm Avenue, Sandiacre, Nottingham NG10 5EJ.
Meetings: The Function Room, Erewash Museum, Anchor Row, Ilkeston. 2nd Tuesday monthly, 7.30 p.m.

Ipswich, Orwell Astronomical Society
Secretary: R. Gooding, 168 Ashcroft Road, Ipswich.
Meetings: Orwell Park Observatory, Nacton, Ipswich. Wednesdays, 8 p.m.

Irish Astronomical Association
Secretary: Terry Moseley (President), 6 Collinbridge Drive, Newtownabbey, Co. Antrim BT36 7SX.
Email: terrymosel@aol.com
Meetings: Ashby Building, Stranmillis Road, Belfast. Alternate Wednesdays, 7.30 p.m.

Irish Astronomical Society
Secretary: James O'Connor, PO Box 2547, Dublin 15, Ireland.
Meetings: Ely House, 8 Ely Place, Dublin 2. 1st and 3rd Monday each month.

Isle of Man Astronomical Society
Secretary: James Martin, Ballaterson Farm, Peel, Isle of Man IM5 3AB.
Email: ballaterson@manx.net
Meetings: Isle of Man Observatory, Foxdale. 1st Thursday of each month, 8 p.m.

Isle of Wight Astronomical Society
Secretary: J. W. Feakins, 1 Hilltop Cottages, High Street, Freshwater, Isle of Wight.
Meetings: Unitarian Church Hall, Newport, Isle of Wight. Monthly.

Keele Astronomical Society
Secretary: Natalie Webb, Department of Physics, University of Keele, Keele, Staffordshire, ST5 5BG.
Meetings: As arranged during term time.

Kettering and District Astronomical Society
Asst. Secretary: Steve Williams, 120 Brickhill Road, Wellingborough, Northants.
Meetings: Quaker Meeting Hall, Northall Street, Kettering, Northants. 1st Tuesday each month, 7.45 p.m.

King's Lynn Amateur Astronomical Association
Secretary: P. Twynman, 17 Poplar Avenue, RAF Marham, King's Lynn.
Meetings: As arranged.

Lancaster and Morecambe Astronomical Society
Secretary: Mrs E. Robinson, 4 Bedford Place, Lancaster, LA1 4EB.
Email: ehelenerob@btinternet.com
Meetings: Church of the Ascension, Torrisholme. 1st Wednesday each month, except July and Aug.

Lancaster University Astronomical Society
Secretary: c/o Students Union, Alexandra Square, University of Lancaster.
Meetings: As arranged.

Laymans Astronomical Society
Secretary: John Evans, 10 Arkwright Walk, The Meadows, Nottingham.
Meetings: The Popular, Bath Street, Ilkeston, Derbyshire. Monthly.

Leeds Astronomical Society
Secretary: Mark A. Simpson, 37 Roper Avenue, Gledhow, Leeds, LS8 1LG.
Meetings: Centenary House, North Street. 2nd Wednesday each month, 7.30 p.m.

Leicester Astronomical Society
Secretary: Dr P. J. Scott, 21 Rembridge Close, Leicester LE3 9AP.
Meetings: Judgemeadow Community College, Marydene Drive, Evington, Leicester. 2nd and 4th Tuesdays each month, 7.30 p.m.

Letchworth and District Astronomical Society
Secretary: Eric Hutton, 14 Folly Close, Hitchin, Herts.
Meetings: As arranged.

Lewes Amateur Astronomers
Secretary: Christa Sutton, 8 Tower Road, Lancing, West Sussex, BN15 9HT.
Meetings: The Bakehouse Studio, Lewes. Last Wednesday each month.

Limerick Astronomy Club
Secretary: Tony O'Hanlon, 26 Ballycannon Heights, Meelick, Co. Clare, Eire.
Meetings: Limerick Senior College, Limerick, Ireland. Monthly (except June and Aug.), 8 p.m.

Lincoln Astronomical Society
Secretary: David Swaey, 'Everglades', 13 Beaufort Close, Lincoln LN2 4SF.
Meetings: The Lecture Hall, off Westcliffe Street, Lincoln. 1st Tuesday each month.

Liverpool Astronomical Society
Secretary: Mr K. Clark, 31 Sandymount Drive, Wallasey, Merseyside L45 0LJ.
Meetings: Lecture Theatre, Liverpool Museum. 3rd Friday each month, 7 p.m.

Norman Lockyer Observatory Society
Secretary: G. E. White, PO Box 9, Sidmouth EX10 0YQ.
Web site: www.ex.ac.uk/nlo/; *Email:* g.e.white@ex.ac.uk
Meetings: Norman Lockyer Observatory, Sidmouth. Fridays and 2nd Monday each month, 7.30 p.m.

Loughton Astronomical Society
Secretary: Charles Munton, 14a Manor Road, Wood Green, London N22 4YJ.
Meetings: 1st Theydon Bois Scout Hall, Loughton Lane, Theydon Bois. Weekly.

Lowestoft and Great Yarmouth Regional Astronomers (LYRA) Society
Secretary: Simon Briggs, 28 Sussex Road, Lowestoft, Suffolk.
Meetings: Community Wing, Kirkley High School, Kirkley Run, Lowestoft. 3rd
Thursday each month, 7.30 p.m.

Luton Astronomical Society
Secretary: Mr G. Mitchell, Putteridge Bury, University of Luton, Hitchin Road,
Luton.
Web site: www.lutonastrosoc.org.uk; *Email:* user998491@aol.com
Meetings: Putteridge Bury, Luton. Last Friday each month, 7.30 p.m.

Lytham St Annes Astronomical Association
Secretary: K. J. Porter, 141 Blackpool Road, Ansdell, Lytham St Annes, Lancs.
Meetings: College of Further Education, Clifton Drive South, Lytham St Annes. 2nd
Wednesday monthly Oct.–June.

Macclesfield Astronomical Society
Secretary: Mr John H. Thomson, 27 Woodbourne Road, Sale, Chesire M33 3SY
Web site: www.g0-evp.demon.co.uk; *Email:* jhandlc@yahoo.com
Meetings: Jodrell Bank Science Centre, Goostrey, Cheshire. 1st Tuesday of every
month, 7 p.m.

Maidenhead Astronomical Society
Secretary: Tim Haymes, Hill Rise, Knowl Hill Common, Knowl Hill, Reading
RG10 9YD.
Meetings: Stubbings Church Hall, near Maidenhead. 1st Friday Sept.–June.

Maidstone Astronomical Society
Secretary: Stephen James, 4 The Cherry Orchard, Haddow, Tonbridge, Kent.
Meetings: Nettlestead Village Hall. 1st Tuesday in the month except July and Aug.,
7.30 p.m.

Manchester Astronomical Society
Secretary: Mr Kevin J. Kilburn FRAS, Godlee Observatory, UMIST, Sackville Street,
Manchester M60 1QD.
Web site: www.u-net.com/ph/mas/; *Email:* kkilburn@globalnet.co.uk
Meetings: At the Godlee Observatory. Thursdays, 7 p.m., except below.
Free Public Lectures: Renold Building UMIST, third Thursday Sept.–Mar., 7.30 p.m.

Mansfield and Sutton Astronomical Society
Secretary: Angus Wright, Sherwood Observatory, Coxmoor Road, Sutton-in-
Ashfield, Nottinghamshire NG17 5LF.
Meetings: Sherwood Observatory, Coxmoor Road. Last Tuesday each month,
7.30 p.m.

Mexborough and Swinton Astronomical Society
Secretary: Mark R. Benton, 14 Sandalwood Rise, Swinton, Mexborough, South
Yorkshire, S64 8PN.
Web site: www.msas.org.uk; *Email:* mark@masas.f9.co.uk
Meetings: Swinton WMC. Thursdays, 7.30 p.m.

Mid-Kent Astronomical Society
Secretary: Peter Bassett, 167 Shakespeare Road, Gillingham, Kent, ME7 5QB.
Meetings: Riverside Country Park, Lower Rainham Road, Gillingham. 2nd and last
Fridays each month, 7.45 p.m.

Milton Keynes Astronomical Society
> *Secretary:* Mike Leggett, 19 Matilda Gardens, Shenley Church End, Milton Keynes, MK5 6HT.
> *Web site:* www.mkas.org.uk; *Email:* mike-pat-leggett@shenley9.fsnet.co.uk
> *Meetings:* Rectory Cottage, Bletchley. Alternate Fridays.

Moray Astronomical Society
> *Secretary:* Richard Pearce, 1 Forsyth Street, Hopeman, Elgin, Moray, Scotland.
> *Meetings:* Village Hall Close, Co. Elgin.

Newbury Amateur Astronomical Society
> *Secretary:* Miss Nicola Evans, 'Romaron', Bunces Lane, Burghfield Common, Reading RG7 3DG.
> *Meetings:* United Reformed Church Hall, Cromwell Place, Newbury. 2nd Friday of month, Sept.–June.

Newcastle-on-Tyne Astronomical Society
> *Secretary:* C. E. Willits, 24 Acomb Avenue, Seaton Delaval, Tyne and Wear.
> *Meetings:* Zoology Lecture Theatre, Newcastle University. Monthly.

North Aston Space & Astronomical Club
> *Secretary:* W. R. Chadburn, 14 Oakdale Road, North Aston, Sheffield.
> *Meetings:* To be notified.

Northamptonshire Natural History Society (Astronomy Section)
> *Secretary:* R. A. Marriott, 24 Thirlestane Road, Northampton NN4 8HD.
> *Email:* ram@hamal.demon.co.uk
> *Meetings:* Humfrey Rooms, Castilian Terrace, Northampton. 2nd and last Mondays, most months, 7.30 p.m.

Northants Amateur Astronomers
> *Secretary:* Mervyn Lloyd, 76 Havelock Street, Kettering, Northamptonshire.
> *Meetings:* 1st and 3rd Tuesdays each month, 7.30 p.m.

North Devon Astronomical Society
> *Secretary:* P. G. Vickery, 12 Broad Park Crescent, Ilfracombe, Devon, EX34 8DX.
> *Meetings:* Methodist Hall, Rhododendron Avenue, Sticklepath, Barnstaple. 1st Wednesday each month, 7.15 p.m.

North Dorset Astronomical Society
> *Secretary:* J. E. M. Coward, The Pharmacy, Stalbridge, Dorset.
> *Meetings:* Charterhay, Stourton, Caundle, Dorset. 2nd Wednesday each month.

North Downs Astronomical Society
> *Secretary:* Martin Akers, 36 Timber Tops, Lordswood, Chatham, Kent, ME5 8XQ.
> *Meetings:* Vigo Village Hall. 3rd Thursday each month. 7.30 p.m.

North-East London Astronomical Society
> *Secretary:* Mr B. Beeston, 38 Abbey Road, Bush Hill Park, Enfield EN1 2QN.
> *Meetings:* Wanstead House, The Green, Wanstead. 3rd Sunday each month (except Aug.), 3 p.m.

North Gwent and District Astronomical Society
> *Secretary:* Jonathan Powell, 14 Lancaster Drive, Gilwern, nr Abergavenny, Monmouthshire, NP7 0AA.
> *Meetings:* Gilwern Community Centre. 15th of each month, 7.30 p.m.

North Staffordshire Astronomical Society
Secretary: Duncan Richardson, Halmerend Hall Farm, Halmerend, Stoke-on-Trent, Staffordshire, ST7 8AW.
Email: dwr@enterprise.net
Meetings: 21st Hartstill Scout Group HQ, Mount Pleasant, Newcastle-under-Lyme ST5 1DR. 1st Tuesday each month (except July and Aug.), 7–9.30 p.m.

North Western Association of Variable Star Observers
Secretary: Jeremy Bullivant, 2 Beaminster Road, Heaton Mersey, Stockport, Cheshire.
Meetings: Four annually.

Norwich Astronomical Society
Secretary: Frank Lawlor, 'Farnworth', Poringland Road, Upper Stoke Holy Cross, Norwich NR14 8NW.
Web site: www.nas.gurney.org.uk
Meetings: Seething Observatory, Toad Lane, Thwaite St Mary, Norfolk. Every Friday, 7.30 p.m.

Nottingham Astronomical Society
Secretary: C. Brennan, 40 Swindon Close, The Vale, Giltbrook, Nottingham NG16 2WD.
Meetings: Djanogly City Technology College, Sherwood Rise (B682). 1st and 3rd Thursdays each month, 7.30 p.m.

Oldham Astronomical Society
Secretary: P. J. Collins, 25 Park Crescent, Chadderton, Oldham.
Meetings: Werneth Park Study Centre, Frederick Street, Oldham. Fortnightly, Friday.

Open University Astronomical Society
Secretary: Dr Andrew Norton, Department of Physics and Astronomy, The Open University, Walton Hall, Milton Keynes MK7 6AA.
Web site: www.physics.open.ac.uk/research/astro/a_club.html
Meetings: Open University, Milton Keynes. 1st Tuesday of every month, 7.30 p.m.

Orpington Astronomical Society
Secretary: Dr Ian Carstairs, 38 Brabourne Rise, Beckenham, Kent BR3 2SG.
Meetings: High Elms Nature Centre, High Elms Country Park, High Elms Road, Farnborough, Kent. 4th Thursday each month, Sept.–July, 7.30 p.m.

Papworth Astronomy Club
Contact: Keith Tritton, Magpie Cottage, Fox Street, Great Gransden, Sandy, Bedfordshire SG19 3AA.
Email: kpt2@tutor.open.ac.uk
Meetings: Bradbury Progression Centre, Church Lane, Papworth Everard, near Huntingdon. 1st Wednesday each month, 7 p.m.

Peterborough Astronomical Society
Secretary: Sheila Thorpe, 6 Cypress Close, Longthorpe, Peterborough.
Meetings: 1st Thursday every month, 7.30 p.m.

Plymouth Astronomical Society
Secretary: Alan G. Penman, 12 St Maurice View, Plympton, Plymouth, Devon PL7 1FQ.
Email: oakmount12@aol.com
Meetings: Glynis Kingham Centre, YMCA Annex, Lockyer Street, Plymouth. 2nd Friday each month, 7.30 p.m.

PONLAF
Secretary: Matthew Hepburn, 6 Court Road, Caterham, Surrey CR3 5RD.
Meetings: Room 5, 6th floor, Tower Block, University of North London. Last Friday
each month during term time, 6.30 p.m.

Port Talbot Astronomical Society (was **Astronomical Society of Wales**)
Secretary: Mr J. Hawes, 15 Lodge Drive, Baglan, Port Talbot, West Glamorgan
SA12 8UD.
Meetings: Port Talbot Arts Centre. 1st Tuesday each month, 7.15 p.m.

Portsmouth Astronomical Society
Secretary: G. B. Bryant, 81 Ringwood Road, Southsea.
Meetings: Monday, fortnightly.

Preston & District Astronomical Society
Secretary: P. Sloane, 77 Ribby Road, Wrea Green, Kirkham, Preston, Lancs.
Meetings: Moor Park (Jeremiah Horrocks) Observatory, Preston. 2nd Wednesday,
last Friday each month, 7.30 p.m.

Reading Astronomical Society
Secretary: Mrs Ruth Sumner, 22 Anson Crescent, Shinfield, Reading RG2 8JT.
Meetings: St Peter's Church Hall, Church Road, Earley. 3rd Friday each month,
7 p.m.

Renfrewshire Astronomical Society
Secretary: Ian Martin, 10 Aitken Road, Hamilton, South Lanarkshire ML3 7YA.
Web site: www.renfrewshire-as.co.uk; *Email:* RenfrewAS@aol.com
Meetings: Coats Observatory, Oakshaw Street, Paisley. Fridays, 7.30 p.m.

Rower Astronomical Society
Secretary: Mary Kelly, Knockatore, The Rower, Thomastown, Co. Kilkenny, Eire.

St Helens Amateur Astronomical Society
Secretary: Carl Dingsdale, 125 Canberra Avenue, Thatto Heath, St Helens,
Merseyside WA9 5RT.
Meetings: As arranged.

Salford Astronomical Society
Secretary: Mrs Kath Redford, 2 Albermarle Road, Swinton, Manchester M27 5ST.
Meetings: The Observatory, Chaseley Road, Salford. Wednesdays.

Salisbury Astronomical Society
Secretary: Mrs R. Collins, 3 Fairview Road, Salisbury, Wiltshire, SP1 1JX.
Meetings: Glebe Hall, Winterbourne Earls, Salisbury. 1st Tuesday each month.

Sandbach Astronomical Society
Secretary: Phil Benson, 8 Gawsworth Drive, Sandbach, Cheshire.
Meetings: Sandbach School, as arranged.

Sawtry & District Astronomical Society
Secretary: Brooke Norton, 2 Newton Road, Sawtry, Huntingdon, Cambridgeshire,
PE17 5UT.
Meetings: Greenfields Cricket Pavilion, Sawtry Fen. Last Friday each month.

Scarborough & District Astronomical Society
Secretary: Mrs S. Anderson, Basin House Farm, Sawdon, Scarborough, N. Yorks.
Meetings: Scarborough Public Library. Last Saturday each month, 7–9 p.m.

Scottish Astronomers Group
Secretary: Dr Ken Mackay, Hayford House, Cambusbarron, Stirling, FK7 9PR.
Meetings: North of Hadrian's Wall, twice yearly.

Sheffield Astronomical Society
Secretary: Mr Andrew Green, 11 Lyons Street, Ellesmere, Sheffield S4 7QS.
Web site: www.saqqara.demon.co.uk/sas/sashome.htm
Meetings: Twice monthly at Mayfield Environmental Education Centre, David Lane, Fulwood, Sheffield S10, 7.30–10 p.m.

Shetland Astronomical Society
Secretary: Peter Kelly, The Glebe, Fetlar, Shetland, ZE2 9DJ.
Email: theglebe@zetnet.co.uk
Meetings: Fetlar, Fridays, Oct.–Mar.

Shropshire Astronomical Society
Secretary: Mrs Jacqui Dodds, 35 Marton Drive, Wellington, Telford, TF1 3HL.
Web site: www.astro.cf.ac.uk/sas/sasmain.html; *Email:* jacquidodds@ntlworld.com
Meetings: Gateway Arts and Education Centre, Chester Street, Shrewsbury.
Occasional Fridays plus monthly observing meetings, Rodington Village Hall.

Sidmouth and District Astronomical Society
Secretary: M. Grant, Salters Meadow, Sidmouth, Devon.
Meetings: Norman Lockyer Observatory, Salcombe Hill. 1st Monday in each month.

Skipton & Craven Astronomical Society
Contact: Tony Ireland, 14 Cross Bank, Skipton, North Yorkshire BD23 6AH.
Email: sacas@andybat.demon.co.uk
Meetings: 3rd Wednesday of each month, Sept.–May.

Solent Amateur Astronomers
Secretary: Ken Medway, 443 Burgess Road, Swaythling, Southampton SO16 3BL.
Web site: www.delscope.demon.co.uk;
Email: kenmedway@kenmedway.demon.co.uk
Meetings: Room 8, Oaklands, Community School, Fairisle Road, Lordshill, Southampton. 3rd Tuesday each month, 7.30 p.m.

Southampton Astronomical Society
Secretary: John Thompson, 4 Heathfield, Hythe, Southampton, SO45 5BJ.
Web site: www.home.clara.net/lmhobbs/sas.html
Email: John.G.Thompson@Tesco.net
Meetings: Conference Room 3, The Civic Centre, Southampton. 2nd Thursday each month (except Aug.), 7.30 p.m.

South Downs Astronomical Society
Secretary: J. Green, 46 Central Avenue, Bognor Regis, West Sussex, PO21 5HH.
Web site: www.southdowns.org.uk
Meetings: Assembly Rooms, Chichester. 1st Friday in each month.

South-East Essex Astronomical Society
Secretary: C. P. Jones, 29 Buller Road, Laindon, Essex.
Web site: www.seeas.dabsol.co.uk/; *Email:* cpj@cix.co.uk
Meetings: Lecture Theatre, Central Library, Victoria Avenue, Southend-on-Sea.
Generally 1st Thursday in month, Sept.–May, 7.30 p.m.

South-East Kent Astronomical Society
Secretary: Andrew McCarthy, 25 St Paul's Way, Sandgate, near Folkestone, Kent, CT20 3NT.
Meetings: Monthly.

South Lincolnshire Astronomical & Geophysical Society
Secretary: Ian Farley, 12 West Road, Bourne, Lincolnshire, PE10 9PS.
Meetings: Adult Education Study Centre, Pinchbeck. 3rd Wednesday each month,
7.30 p.m.

Southport Astronomical Society
Secretary: Patrick Brannon, Willow Cottage, 90 Jacksmere Lane, Scarisbrick,
Ormskirk, Lancashire, L40 9RS.
Meetings: Monthly Sept.–May, plus observing sessions.

Southport, Ormskirk and District Astronomical Society
Secretary: J. T. Harrison, 92 Cottage Lane, Ormskirk, Lancs L39 3NJ.
Meetings: Saturday evenings, monthly as arranged.

South Shields Astronomical Society
Secretary: c/o South Tyneside College, St George's Avenue, South Shields.
Meetings: Marine and Technical College. Each Thursday, 7.30 p.m.

South Somerset Astronomical Society
Secretary: G. McNelly, 11 Laxton Close, Taunton, Somerset.
Meetings: Victoria Inn, Skittle Alley, East Reach, Taunton, Somerset. Last Saturday
each month, 7.30 p.m.

South-West Hertfordshire Astronomical Society
Secretary: Tom Walsh, 'Finches', Coleshill Lane, Winchmore Hill, Amersham,
Buckinghamshire HP7 0NP.
Meetings: Rickmansworth. Last Friday each month, Sept.–May.

Stafford and District Astronomical Society
Secretary: Miss L. Hodkinson, 6 Elm Walk, Penkridge, Staffordshire, ST19 5NL.
Meetings: Weston Road High School, Stafford. Every 3rd Thursday, Sept.–May,
7.15 p.m.

Stirling Astronomical Society
Secretary: Hamish MacPhee, 10 Causewayhead Road, Stirling FK9 5ER.
Meetings: Smith Museum & Art Gallery, Dumbarton Road, Stirling. 2nd Friday each
month, 7.30 p.m.

Stoke-on-Trent Astronomical Society
Secretary: M. Pace, Sundale, Dunnocksfold, Alsager, Stoke-on-Trent.
Meetings: Cartwright House, Broad Street, Hanley. Monthly.

Stratford-upon-Avon Astronomical Society
Secretary: Robin Swinbourne, 18 Old Milverton, Leamington Spa, Warwickshire,
CV32 6SA.
Meetings: Tiddington Home Guard Club. 4th Tuesday each month, 7.30 p.m.

Sunderland Astronomical Society
Contact: Don Simpson, 78 Stratford Avenue, Grangetown, Sunderland SR2 8RZ.
Meetings: Friends Meeting House, Roker. 1st, 2nd and 3rd Sundays each month.

Sussex Astronomical Society
Secretary: Mrs C. G. Sutton, 75 Vale Road, Portslade, Sussex.
Meetings: English Language Centre, Third Avenue, Hove. Every Wednesday,
7.30–9.30 p.m., Sept.–May.

Swansea Astronomical Society
Secretary: Dr Michael Morales, 238 Heol Dulais, Birch Grove, Swansea SA7 9LH.
Web site: www.crysania.co.uk/sas/astro/star
Meetings: Lecture Room C, Science Tower, University of Swansea. 2nd and 4th
Thursday each month from September to June, 7 p.m.

Tavistock Astronomical Society
Secretary: Mrs Ellie Coombes, Rosemount, Under Road, Gunnislake, Cornwall PL18 9JL.
Meetings: Science Laboratory, Kelly College, Tavistock. 1st Wednesday each month, 7.30 p.m.

Thames Valley Astronomical Group
Secretary: K. J. Pallet, 82a Tennyson Street, South Lambeth, London SW8 3TH.
Meetings: As arranged.

Thanet Amateur Astronomical Society
Secretary: P. F. Jordan, 85 Crescent Road, Ramsgate.
Meetings: Hilderstone House, Broadstairs, Kent. Monthly.

Torbay Astronomical Society
Secretary: Tim Moffat, 31 Netley Road, Newton Abbot, Devon, TQ12 2LL.
Meetings: Torquay Boys' Grammar School, 1st Thursday in month; and Town Hall, Torquay, 3rd Thursday in month, Oct.–May, 7.30 p.m.

Tullamore Astronomical Society
Secretary: Tom Walsh, 25 Harbour Walk, Tullamore, Co. Offaly, Eire.
Web site: www.iol.ie/seanmck/tas.htm; *Email:* tcwalsh25@yahoo.co.uk
Meetings: Order of Malta Lecture Hall, Tanyard, Tullamore, Co. Offaly, Eire. Mondays at 8 p.m., every fortnight.

Tyrone Astronomical Society
Secretary: John Ryan, 105 Coolnafranky Park, Cookstown, Co. Tyrone.
Meetings: Contact Secretary.

Usk Astronomical Society
Secretary: Bob Wright, 'Llwyn Celyn', 75 Woodland Road, Croesyceiliog, Cwmbran, NP44 2OX.
Meetings: Usk Community Education Centre, Maryport Street, Usk. Every Thursday during school term, 7 p.m.

Vectis Astronomical Society
Secretary: Rosemary Pears, 1 Rockmount Cottages, Undercliff Drive, St Lawrence, Ventnor, Isle of Wight PO38 1XG.
Web site: www.wightskies.fsnet.co.uk/main.html
Email: may@tatemma.freeserve.co.uk
Meetings: Lord Louis Library Meeting Room, Newport. 4th Friday each month except Dec., 7.30 p.m.

Vigo Astronomical Society
Secretary: Robert Wilson, 43 Admers Wood, Vigo Village, Meopham, Kent DA13 0SP.
Meetings: Vigo Village Hall. As arranged.

Walsall Astronomical Society
Secretary: Bob Cleverley, 40 Mayfield Road, Sutton Coldfield, B74 3PZ.
Meetings: Freetrade Inn, Wood Lane, Pelsall North Common. Every Thursday.

Wellingborough District Astronomical Society
Secretary: S. M. Williams, 120 Brickhill Road, Wellingborough, Northants.
Meetings: Gloucester Hall, Church Street, Wellingborough. 2nd Wednesday each month, 7.30 p.m.

Wessex Astronomical Society
Secretary: Leslie Fry, 14 Hanhum Road, Corfe Mullen, Dorset.
Meetings: Allendale Centre, Wimborne, Dorset. 1st Tuesday of each month.

West Cornwall Astronomical Society

Secretary: Dr R. Waddling, The Pines, Pennance Road, Falmouth, Cornwall TR11 4ED.

Meetings: Helston Football Club, 3rd Thursday each month, and St Michalls Hotel, 1st Wednesday each month, 7.30 p.m.

West of London Astronomical Society

Secretary: Duncan Radbourne, 28 Tavistock Road, Edgware, Middlesex HA8 6DA.

Web site: www.wocas.org.uk

Meetings: Monthly, alternately in Uxbridge and North Harrow. 2nd Monday in month, except Aug.

West Midlands Astronomical Association

Secretary: Miss S. Bundy, 93 Greenridge Road, Handsworth Wood, Birmingham.

Meetings: Dr Johnson House, Bull Street, Birmingham. As arranged.

West Yorkshire Astronomical Society

Secretary: Pete Lunn, 21 Crawford Drive, Wakefield, West Yorkshire.

Meetings: Rosse Observatory, Carleton Community Centre, Carleton Road, Pontefract. Each Tuesday, 7.15 p.m.

Whitby and District Astronomical Society

Secretary: Rosemary Bowman, The Cottage, Larpool Drive, Whitby, North Yorkshire, YO22 4ND.

Meetings: Whitby Mission, Seafarer's Centre, Haggersgate, Whitby. 1st Tuesday of the month, 7.30 p.m.

Whittington Astronomical Society

Secretary: Peter Williamson, The Observatory, Top Street, Whittington, Shropshire.

Meetings: The Observatory. Every month.

Wiltshire Astronomical Society

Secretary: Simon Barnes, 25 Woodcombe, Melksham, Wilts SN12 6HA.

Meetings: St Andrews Church Hall, Church Lane, off Forest Road, Melksham, Wilts.

Wolverhampton Astronomical Society

Secretary: Mr M. Bryce, Iona, 16 Yellowhammer Court, Kidderminster, Worcestershire, DY10 4RR.

Web site: www.wolvas.org.uk; *Email:* michaelbryce@wolvas.org.uk

Meetings: Beckminster Methodist Church Hall, Birches Barn Road, Wolverhampton. Alternate Mondays, Sept.–Apr., extra dates in summer, 7.30 p.m.

Worcester Astronomical Society

Secretary: Mr S. Bateman, 12 Bozward Street, Worcester WR2 5DE.

Meetings: Room 117, Worcester College of Higher Education, Henwick Grove, Worcester. 2nd Thursday each month, 8 p.m.

Worthing Astronomical Society

Contact: G. Boots, 101 Ardingly Drive, Worthing, West Sussex, BN12 4TW.

Web site: www.worthingastro.freeserve.co.uk

Email: gboots@observatory99.freeserve.co.uk

Meetings: Heene Church Rooms, Heene Road, Worthing. 1st Wednesday each month (except Aug.), 7.30 p.m.

Wycombe Astronomical Society

Secretary: Mr P. Treherne, 34 Honeysuckle Road, Widmer End, High Wycombe, Buckinghamshire, HP15 6BW.

Meetings: Woodrow High House, Amersham. 3rd Wednesday each month, 7.45 p.m.

The York Astronomical Society
  Contact: Hazel Collett, Public Relations Officer
  Telephone: 07944 751277
  Web site: www.yorkastro.freeserve.co.uk; *Email:* info@yorkastro.co.uk
  Meetings: The Knavesmire Room, York Priory Street Centre, Priory Street, York.
  1st and 3rd Friday of each month (except Aug.), 8 p.m.

Any society wishing to be included in this list of local societies or to update details, including any web-site addresses, is invited to write to the Editor (c/o Pan Macmillan, 20 New Wharf Road, London N1 9RR), so that the relevant information may be included in the next edition of the *Yearbook*.

The William Herschel Society maintains the museum established at 19 New King Street, Bath BA1 2BL – the only surviving Herschel House. It also undertakes activities of various kinds. New members would be welcome; those interested are asked to contact the Membership Secretary at the museum.

The South Downs Planetarium (Kingsham Farm, Kingsham Road, Chichester, West Sussex PO19 8RP) is now fully operational. For further information, visit www.southdowns.org.uk/sdpt or telephone (01243) 774400